AF730618

Table of Contents

1. LABORATORY ETIQUETTE AND SAFETY

Your laboratory training will imbue skills in planning, organizing, and problem solving that will benefit every aspect of your life.
Laboratory training prepares you for jobs in the public and private sectors. There are many high paying technician positions in research labs at universities, government institutions, and pharmaceutical, biotechnology, and medical labs. Lab training is essential for a career in academia.

A. *Etiquette*

Laboratory research and routine laboratory analysis can be stressful. Experiments fail, cultures become contaminated, and so on. Almost all laboratory work involves teamwork and collaboration. To minimize the stress and maintain good work relationships, it is helpful to accept and hold to practices that have been shown to positively influence outcomes.
Good lab practices include:

- Checking your anger and personal problems at the door. Be cordial and kind to your teammates; nothing is gained from harsh words or name-calling
- Doing your share of the work if you are part of a team
- Communicating regularly and promptly
- Sharing information regularly with your team
- Encouraging openness and diversity
- Being creative and observant – new ideas and findings move science forward
- Asking permission before using another person's materials (reagents, cells, pipettor, *etc.*)
- Cleaning up your own mess
- Telling the lab manager or Principal Investigator (PI) about broken equipment or supplies that need replenishing

> Some people like to lead. A good leader works to engage everyone on the team and respects diverse ideas. A diverse team might include people of different age, gender, race, culture, sexual preference, and viewpoints. Groups composed of people with diverse experiences and areas of expertise tend to be more creative and innovative.

B. *Self-improvement and lab discipline*

Lab work can help you learn discipline and improve your work habits.

- **Set specific, realistic goals for self-improvement**

Example	Identify the improvement or goal you seek	Break the goal into small, manageable increments
1	*I want to create a poster for the undergraduate research competition at the end of the semester; it would be nice to win a prize for my poster*	*I will plan three experiments – one per week; after each experiment, I will work with my mentor to prepare a figure that explains the data and can be included on the poster*
2	*I want to get a grade of A or B in each of my classes this semester.*	*I will set up a calendar and use it to keep track of the assignments due in each class. I will include a checkbox to mark when an assignment is complete. I learned that keeping up with assignments and looking at material regularly is a great way to enhance learning and improve grades.*

- **Motivation** Regularly remind yourself why you are doing "this" ("this" refers to your academic studies, exercise program, volunteer work, bathroom remodeling project *etc.*)
- **Reward** yourself when you complete an important goal
- **Develop good habits** Nike coined the phrase "Just do it" but in reality, this approach does not work well for most individuals. Instead, if you want to make improvements in your life or accomplish a particular goal, you are more likely to succeed if you try to make incremental changes in your routine until an action becomes a habit. It also helps to reduce the barriers to success –
- **Mise en place** means "setting up" or "putting things in place". Chefs use this term to describe best practices for preparing a meal; measure out all of your ingredients before you begin cooking. Employ *mise en place* when planning the layout and storage of materials in your lab and having all the equipment and reagents you need for an experiment laid out on the lab bench before you begin an experiment. Prepare your own reagents if possible. As a general rule, it is not good to borrow from others; never borrow without permission.
- **Organize your learning/creative experience**
  - **File your work** You can file work into folders on a computer or in a few paper folders. It helps to do advance planning: "How should I group the material?" "Will this organization scheme allow me to quickly retrieve a file? For a laboratory class, your files might be organized by week or by topic (handouts, research notes, data analysis, figures, quizzes)
  - **Use specific file names** You may have more than one homework assignment. The file name Bio homeworkTW 02.14.22.docx is more specific than Bio homework.docx
  - **Make lists** Studies show that routine use of lists saves time and improves efficiency of short-term (today's to-do list) and long-term projects (list of experiments and writing steps).

C. *Safety – Personal protection*

- **Skin** Lab coats protect against chemical splashes and spill hazards. Regular clothing should always provide adequate coverage, minimizing skin exposure.
- **Feet** Shoes must fully cover the feet. Most labs do not permit open-toe shoes or sandals.
- **Eyes** Goggles provide the <u>best</u> protection against chemical splashes and impact hazards. Eye and safety glasses are fine but provide less protection due to their open frame design. Contact lens do not provide protection and can trap hazardous chemicals that damage delicate eye tissue. Therefore, goggles or safety glasses <u>must be worn over contact lenses</u> when working with toxic chemicals.
- **Hands** Protective gloves should always be worn when handling hazardous materials. Choose a glove material that provides barrier against the specific substance in use.
- **Latex gloves** generally resist acids, alkalis, salts, ketones and biohazards (body fluids, bacteria).
- **Nitrile gloves** resist alkaline solutions, most organic solvents (ethanol, hexane, formaldehyde, heptane, petroleum ether, methanol, 2-propanol) and biohazards. <u>Note:</u> cases of <u>allergic reaction</u> to latex and/or glove powders have been reported. Alternative glove materials are available upon request, see instructor.

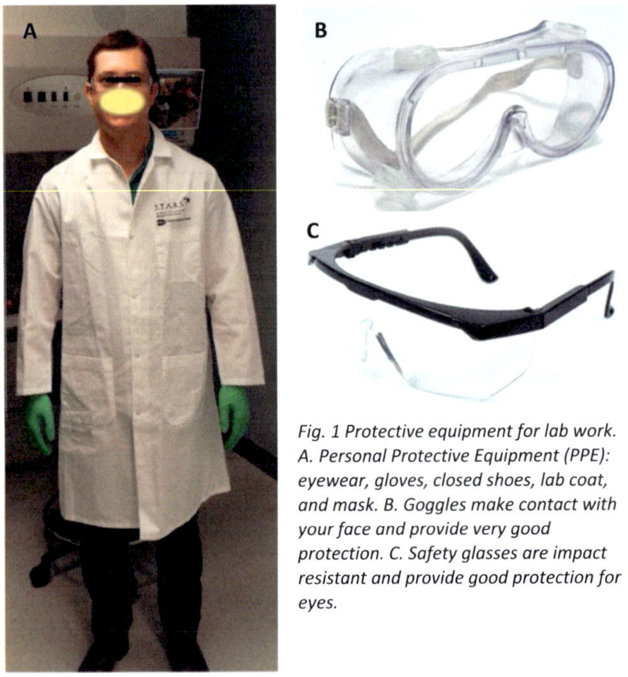

Fig. 1 Protective equipment for lab work. A. Personal Protective Equipment (PPE): eyewear, gloves, closed shoes, lab coat, and mask. B. Goggles make contact with your face and provide very good protection. C. Safety glasses are impact resistant and provide good protection for eyes.

D. *Safety equipment found in most laboratories*

- First aid kit (Fig 2)
- Safety showers (Fig 2)
- Eyewash station
- Fume hood (Fig 3)
- Fire extinguishers (Fig 3)
- Shunt trip alarm
- Fire alarm

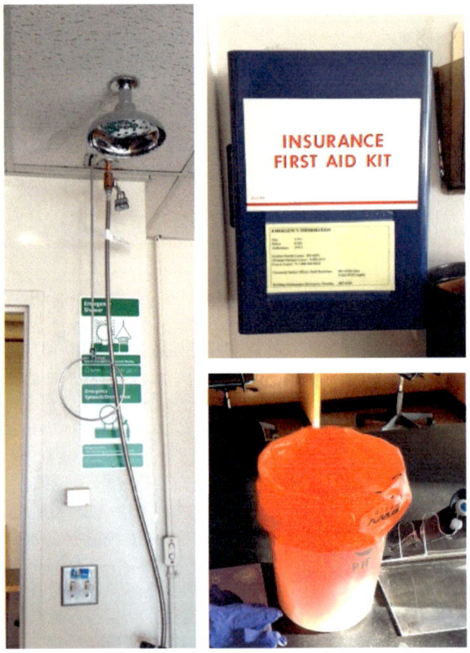

Fig 2. Safety shower (left), First aid kit (upper right) and biowaste container (lower right)

E. *Safe laboratory practices*

- Keep work area clean and uncluttered
- Do not sit on lab bench tops – clothing is thin and fecal microbes can easily pass-through layers of clothing and get deposited on the bench surface.
- Label all solutions with accurate content description. Clean up spills promptly according to approved safety procedures.
- Never eat, drink, smoke, or apply make-up or lotions in the laboratory.
- Remove gloves and wash hands before leaving laboratory area.
- Never mouth-pipet chemicals or directly smell chemicals in open bottles, this can cause severe tissue damage to mouth, throat, and/or lungs.
- Operate lab equipment according to manufacturer's instructions. For example, centrifuges should <u>always</u> be loaded with samples balanced and opposite to each other and rotor lid secured.
- **ONLY** use **Flammable Material Storage Refrigerators** for flammable chemical storage. Household-type units are not safe for flammables, internal sparks may ignite explosions.
- Avoid blocking emergency exits or access to safety equipment with backpacks, etc.

F. *Approved safe waste storage and disposal procedures*

- **Hazardous chemicals**: Store in a clearly labeled Waste Container in the hood with compatible chemicals.
- **Biohazard waste** (Fig 2)**:** Dispose in designated Biohazard Waste Bags. Autoclave or burn bags according to approved biohazard disposal procedures.
- **Sharp Disposal** (Fig 4)**:** Dispose of needles and razor blades in Biohazard Sharps Container. Broken glass can be placed in the Broken Glass Disposal Box, NOT in general use trashcans.
- **Petri Plate Waste**: Put in designated Biohazard Waste Bag.
- **Pipet Tips**: Place tips in Tip Disposal Containers. *Do NOT put tips into regular trash cans.*
- **Preserved Specimens**: Dispose in labeled Preserved Specimen Waste Can. **Always** remove pins and sharps from specimens before placing in can.

Fig 3. Fire extinguisher and fume (chemical) hood

G. *Emergency procedures*

- **Personal Injury**: For minor cuts or puncture wounds: Wash wound with soap and water, apply antiseptic and bandage. With more serious wounds, apply pressure to reduce bleeding and seek medical attention immediately. Always wear gloves to treat an injured person when blood or body fluids are involved.
- **Chemical Spills**: If you splash chemicals on skin or clothing, remove contaminated clothing immediately and flush affected body areas with water for 15 minutes using the nearest eyewash station and/or safety shower. Seek medical attention at nearest medical facility or call 911.
- **Fire**: Use the "stop, drop, and roll" method and/or safety shower to extinguish fire on clothing or hair. Minor fires can be quenched with an appropriately rated fire extinguisher-- pull lock ring out, hold unit upright and position 8 ft. from fire, aim at base of fire, and sweep side to side. Sand is also effective in quenching fires. Otherwise, activate nearest fire alarm, evacuate building, and call 911.
- **Electrical Hazards** (Fig 4): In case of an electrocution or an explosive vapor hazard, push the red SHUNT TRIP BUTTON (if present)– this shuts off room electricity. *Never touch a person* in contact with an electrical current. WAIT for the current to be turned off and call 911.

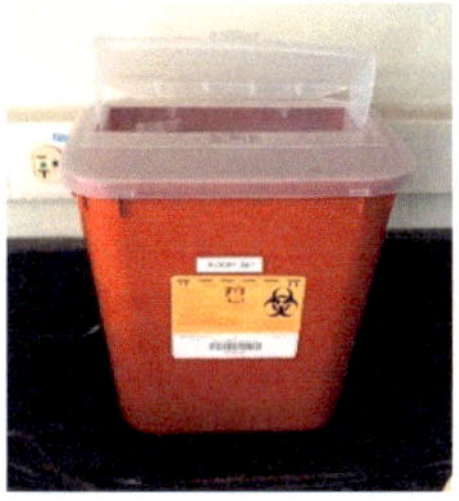

Fig 4. Sharps container

H. *Evacuation procedures*

- If possible, turn off any equipment that might pose a safety hazard for first responders.
- Use the nearest evacuation exit to leave the building. Never use elevators during an emergency evacuation. Stand at the designated waiting area, away from the building, and avoid blocking emergency personnel and vehicles.
- Wait until the emergency responders give the "all clear" signal before re-entering the building.
- Evacuation Procedure for the Disabled: Evacuate disabled person to the nearest Evacuation Assistance Area, denoted by blue and white disable sign.

Fig 5. Many labs have a red SHUNT TRIP button that can be used to shut off electricity to the lab.

2. SCIENTIFIC METHOD AND SCIENCE NOMENCLATURE

A. *The scientific method*

The **scientific method** is a systematic process of investigation that is used to discover how things work.

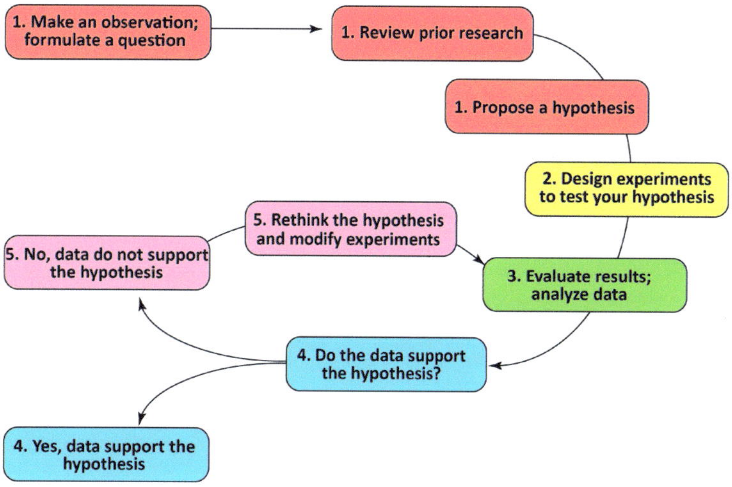

PROCESS STEP 1 (red): The investigator making the **observation** will **formulate a question** and **determine what is known about the topic** (review prior peer-reviewed research). S/he then proposes a **hypothesis** to explain the underlying cause of the observed phenomenon. A hypothesis can be tested.

EXAMPLE

Greek philosophers, including Aristotle, had a theory (an idea to explain observations or events) about gravity. They believed that a heavy object would fall to the ground faster than a light object. Galileo, who is considered the designer of the scientific method, did not agree with the Greeks' theory. Galileo observations led him to believe that objects fall to the earth ***at the same rate***, independent of mass (weight). [Recall that Galileo, a math professor, also believed that the Earth and other planets revolved around the sun. He was ridiculed, imprisoned, and eventually excommunicated from the Catholic church because of his ideas. This is just one example of science denial. It took 400 years for the Catholic church to reverse their decision].

PROCESS STEP 2 (yellow): Experiments to test the hypothesis are planned and executed.

EXAMPLE

In the late 1500's Galileo dropped two balls of very different mass off the leaning tower of Pisa.

PROCESS STEP 3 (green): The experimental results are evaluated.

EXAMPLE

An observer at the base of the Tower of Pisa found that the balls landed on the ground at precisely the same time. The experiment was repeated with the same results.

PROCESS STEP 4 (blue): If the hypothesis aligns with (supports) the predicted outcome, the hypothesis is supported. If the results do not align with the predicted outcome, the investigator takes these results into account and uses this new information to formulate a new, revised, testable hypothesis (Step 5).

EXAMPLE

Galileo concluded that the results supported his hypothesis.

B. *Variables*

The value measured for the **dependent variable** is directly affected by the **independent variable**. Measuring the dependent variable indicates whether the independent variable affects the outcome of the experiment.

EXAMPLE:

In the Galileo example above, the **independent variable** is the mass of the two balls. The **dependent variable** is the landing time.

> The **dependent variable** is the information that is **determined** (or measured) by the investigator. The **independent variable** is the condition that is **influenced** (or changed) by the investigator.

C. *Controls are essential in experimental design*

Controls ensure that the experiment is working the way it was expected to work. The **control group(s)** are objects (i.e. plants, animals, microbes, metal balls) in the experiment that are manipulated in parallel alongside the test objects *but are not changed with respect the independent variable.*

In the controls, all other, unmeasured variables are held constant. If the non-measured variables are not held constant, they may affect the outcome (results). The control treatment minimizes effects of unmeasured variables.

EXAMPLE:

A researcher wants to measure the toxicity of a drug by treating mice with the drug and measuring the number of surviving mice.

- The cohort of mice is divided into two same-size groups: (1) control and (2) treatment (test) group.
- The control group will not receive the drug but will otherwise receive treatment (same food, water, housing, *etc.*) as that of the test group.
- The results for the control will reveal the death rate of the mice *without* drug treatment. This will provide a solid baseline for comparison with the test group.

Generally, the more 'controlled' an experiment is, the more reliable the results are.

Contemplate the role of the control in this simple experiment:

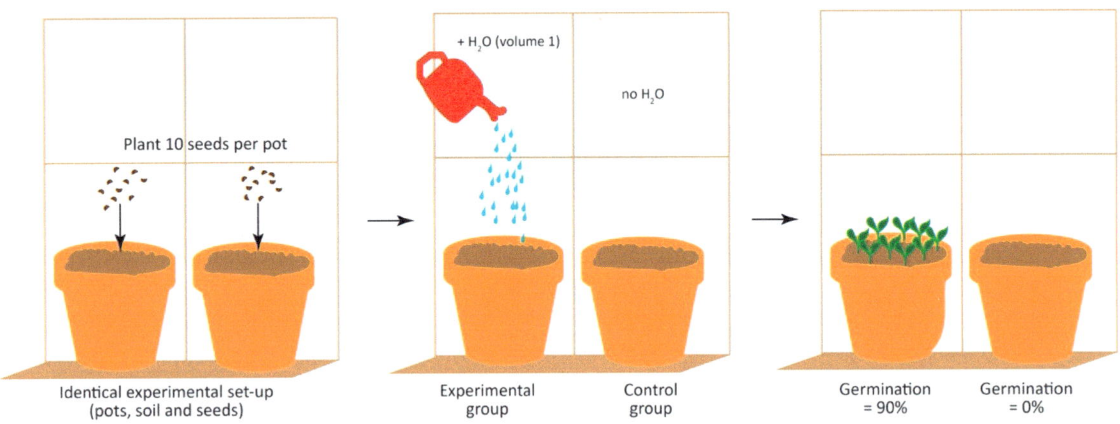

Fig 1. Is water critical for seed germination and growth? In this experiment, an equal number of seeds is planted in two pots. The two pots are given identical treatement except for one variable – the amount of water. One pot is watered and seeds germinate. The second pot, the control, does not receive water. Absense of growth in the second pot supports the conclusion that water is the variable that allowed seeds to germinate.

Practice 1

You study chemical X to determine if it inhibits the germination of petunia seeds. You set up two germination assays on filter paper in sterile Petri dishes. To start the germination process, you add the same volume of aqueous solutions of chemical X (a low and a high concentration) and incubate the seeds in a drawer in the dark for one week.

After one week, you find that none of the seeds have germinated. You conclude the following: **Chemical X inhibits petunia seed germination at low and at high concentrations.** Is the conclusion valid? Explain your answer.

You repeat the experiment but decide to include a control treatment, *i.e.*, addition of water only (no chemical X) to a third set of seeds, in addition to the two treatments you used previously (a low and a high concentration of chemical X). You anticipate that the seeds exposed to the control treatment (water only) will germinate at some basic rate if the system is working properly.

After one week you find open the drawer and find that the rate of germination is very low again. Only 1 out of 10 seeds has germinated in the water treatment and as before, none of the seeds germinated in the seeds treated with chemical X.

Based on these results, how valid are the statements below?

	Valid	*Not valid*
Chemical X inhibits petunia seed germination at low and at high concentrations.		
Petunia seeds do not germinate in the presence of water.		
The petunia seeds that were used have very low germination rates.		
Water alone is not enough to aid in Petunia seed germination.		

Other modifications of the experiment were tested next:

- You test a different batch of petunia seeds in addition to the previous batch to determine if there is something wrong with the seeds.
- You include a "dry" (no water) control
- You decide to place some seeds in the light during the week of incubation to see if this helps with germination.

Here are the germination results after one week.

	No water (dry)		With water		Chemical X low conc.		Chemical X high conc.	
	Dark	**Light**	**Dark**	**Light**	**Dark**	**Light**	**Dark**	**Light**
New seeds	0/10	0/10	0/10	9/10	0/10	8/10	0/10	5/10
Old seeds	0/10	0/10	1/10	9/10	0/10	9/10	0/10	4/10

What conclusions can you draw from these data?

Why was setting up a control experiment (water only) important for the successful completion of the experiment?

D. *Nomenclature in biology*

Nomenclature (name calling) [pronunciation: no men clay tur; from dictionary: nōmən ˌklāCHər] is the system used to name things. Mnemonic: nomen = name. Life scientists use the Linnean binomial system (bi means *two*) with two names — *genus* <u>and</u> *species* — to identify an organism.

Genus and species are always *italicized* or <u>underlined.</u>

Nomenclature rules

1. The genus name is capitalized and italicized or underlined. The genus is a more general taxonomic category than is species. Often members of the same genus have similarities in appearance. For example, you can recognize different species of bears (genus *Ursa*) by their appearance.
2. The species name is all lower-case letters and italicized or underlined. A species is a group of organisms with similar anatomical characteristics. In the case of organisms that undergo sexual reproduction, species members can interbreed to produce fertile offspring. In general, different species do not tend to interbreed, but there are examples of hybrids in nature (for example, certain birds) that result when separate species breed successfully. Some hybrids are fertile; others (such as mules) cannot reproduce (they are sterile). Cauliflower and kale are the same genus (*Brassica*) and species (*oleracea*) but they are very different in appearance — they are examples of cultivars (not italicized). In contrast, *Canis lupus* is the genus species name for ≈ eight extant (existing) animals including the domestic dog and the grey wolf. With the advent of sequencing, the concept of species in bacteria has been evolving. Since bacteria are not capable of sexual reproduction, scientists compare DNA sequences and traits (phenotypes) to distinguish bacterial species.

EXAMPLES

	Genus	species
humans	*Homo*	*sapiens*
humanoid (extinct)	*Homo*	*erectus*
cycad (gymnosperm)	*Cycas*	*rumphii*
cauliflower	*Brassica*	*oleracea*
kale	*Brassica*	*oleracea* (Acephala Group)
dog	*Canis*	*lupus (subspecies familiaris)*
grey wolf	*Canis*	*lupus*
domestic cat	*Felis*	*catus*
mountain lion	*Puma*	*concolor*

Practice 2

Give the genus species name for the common fruit fly: ______________________________

Give the genus species name for baker's yeast: ______________________________

3. METRIC REVIEW

The metric system is very 'user friendly'. Unfortunately, the USA (and Liberia and Myanmar) still uses the Imperial System, which we inherited from the British Empire >200 years ago. The Imperial System uses a set of unrelated measures and is not a 'user-friendly' system. Let's make America metric!

The French Revolution not only toppled a king—it also forged the metric system

As revolution raged in the 1790s, French scientists replaced a chaotic system of weights and measures with a unified way to calibrate and calculate.

Fig. 2 Image from Atlas Obscura atlasobscura.com

A. *Metric basics*

Scientists and medical professionals use the metric system. Here are useful features of the metric system.

- based on multiples of 10
- prefixes are used to designate levels (powers of 10, such as kilo, which means 1000 or 10^3); prefixes are consistent throughout the metric system, which greatly simplifies learning
- the difference between the most prefixes is <u>1000</u> (exception centi- as in cm = 1/100 of a meter)

Measure	Base unit	Abbr.
length	meter	m
weight (mass)	gram	g
volume	liter	l

Prefixes

Use a prefix to <u>decrease</u> the base

	Prefix (examples)		
Base (abbreviation)	**centi: 1/100 or 10^{-2}**	**milli: 1/1000 or 10^{-3}**	**micro: 1/1,000,000 or 10^{-6}**
meter (m)	**centi**meter (cm)	**milli**meter (mm)	**micro**meter(µm)
gram (g)		**milli**gram (mg)	**micro**gram (µg)
liter (l)		**milli**liter (ml)	**micro**liter (µl)

Use a prefix to <u>increase</u> the base

	Prefix (examples)
Base (abbreviation)	**kilo: 1000 (10^3)**
meter (m)	**kilo**meter (km)
gram (g)	**kilo**gram (kg)

The metric system: prefixes used commonly in biology (the BASE unit is shown in red)

The example below is for the base **meter.**

Prefix	example	abbr.	this	equals this in decimal form	scientific notation
tera	terameter	Tm	1 Tm	1,000,000,000,000.0 meter	10^{12}
giga	gigameter	Gm	1 Gm	1,000,000,000.0 meter	10^{9}
mega	megameter	Mm	1 Mm	1,000,000.0 meter	10^{6}
kilo	kilometer	km	1 km	1000.0 meter	10^{3}
-----	meter	m	1 m	1.0 meter	1
centi	centimeter	cm	1 cm	0.01 meter	10^{-2}
milli	millimeter	mm	1 mm	0.001 meter	10^{-3}
micro	micrometer	µm	1 µm	0.000001 meter	10^{-6}
nano	nanometer	nm	1 nm	0.000000001 meter	10^{-9}
pico	picometer	pm	1 pm	0.000000000001 meter	10^{-12}

B. *Scientific notation (SCI on some calculators)*

Metric line
(partial list)

10^{12}	10^{9}	10^{6}	10^{3}	10^{0}	10^{-3}	10^{-6}	10^{-9}	10^{-12}
Terra	Giga	Mega	kilo	base	milli	micro	nano	pico
T	G	M	k	(gram) (liter) (meter)	m	µ	n	p

000000000000001.00000000000000

Scientific notation is used to express very large or very small numbers.

EXAMPLE

1,000,000 (1 million)	$= 1.0 \times 10^{6}$
45,000,000 (45 million)	$= 4.5 \times 10^{7}$
0.000056	$= 5.6 \times 10^{-5}$

Conversions:
When converting a measurement with a large unit, such as kilogram, to a smaller unit, such as gram, the *number* will INCREASE.
Example: 32 kg = 32000 g
When converting a measurement with a small unit, such as milligram, to a larger unit, such as gram, the *number* will DECREASE.
the unit increases (mg to g)
Example: 14 mg = 0.014 g
the number decreases (14 to 0.014)

Practice 1:
Rewrite the numbers below using scientific notation:

1,000,000	= ________________
265,000,000.000	= ________________
34,000	= ________________
0.000045	= ________________
0.0000006789	= ________________

For numbers less than 1, add a zero before the decimal point:
CORRECT 0.13
INCORRECT .13

C. *Simple conversions*

Since 1 mg = 0.001 gram, then 1 g = 1000 mg [math: multiply both sides by 1000]
Since 1 µl = 10^{-6} liter, then 1 liter = 1,000,000 µl [math: multiply both sides by 1,000,000 or 10^{6}]

Practice 2:

Convert			**to**
X	unit	=	unit
100	mg		g
0.3	g		mg
205	mg		g
51	ml		l
0.051	l		ml

D. *Conversions and scientific notation*

Practice 3:
Volume conversions

Convert		**Write in scientific notation**
9 liter is equivalent to	9000 ml	9.0×10^{3} ml
900 l is equivalent to	______ ml	________ ml
0.9 l is equivalent to	______ µl	________ µl
90 ml is equivalent to	______ µl	________ µl

Weight conversions

Convert		**Write in scientific notation**
1 kg is equivalent to	______ mg	________________
1 g is equivalent to	______ mg	________________
1 g is equivalent to	______ µg	________________
0.324 µg is equivalent to	______ g	________________
0.005 g is equivalent to	______ µg	________________

E. *Significant figures*

There are several rules to remember when determining the number of significant figures

1. Non-zero digits are <u>always</u> significant.
2. Any zeros <u>between</u> two significant digits are significant.
3. A final zero or trailing zeros after numbers and after the decimal point are significant.
4. Zeros after the decimal point and before non-zero numbers are <u>not</u> significant.

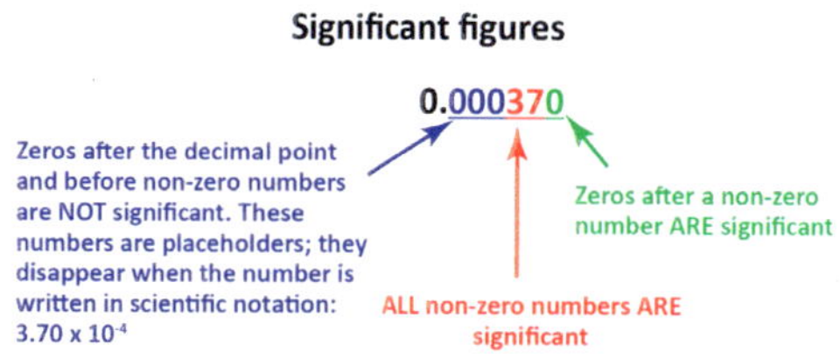

Remember this rule: **your answer can <u>not</u> be MORE precise than the least precise measurement.**

EXAMPLE

How many significant figures are in the answer to this calculation: **2.33 x 6.085 x 2.1**?

The number of significant figures will be <u>two</u> because the *least precise measurement*, 2.1, has two significant numbers.

More about zeros:

Examples: 0.**00**500 and 0.**0**3040 → The zeros in bold serve only as space holders. They are there to put the decimal point in its correct location. They DO NOT involve measurement decisions. <u>Upon writing the numbers in scientific notation (5.00×10^{-3} and 3.040×10^{-2}), the non-significant zeros disappear.</u>

An **<u>exact number</u>** does not affect the number of significant figures in a calculation. An exact number is a value that is known with complete certainty – numbers that are definitions, such as a dozen (12) or when referring to a conversion (100 cm in 1 m, 60 sec in 1 min). Exact numbers, such as the number of people in a room, have an infinite number of significant figures. Exact numbers are numbers that are exact by definition, such as: 1 inch = 2.54 cm or 1 gallon = 231 cubic inches or 1 foot = 12 inches or 1 meter = 100 centimeters or numbers that come in integers and are not likely to be available in amounts smaller than integers. (When you ask for seating in a restaurant, the number of people is an integer, an exact number). Exact numbers are not measurements made with instruments, such as a balance.

If a number is exact, it DOES NOT affect the accuracy of a calculation nor the precision of the expression.

EXAMPLE

How many significant figures are in the answer to this calculation? **32650 cookies x 12 (as in dozen)**

The number of significant figures will be <u>four</u> because **32650** has four significant numbers and we can ignore 12 because *it is an exact number*.

32650 cookies x 12 = 2720.83333 = 2721 dozen cookies

Don't rely on your calculator; it does not know about significant numbers — you must trim non-significant numbers.

Here are some examples:

Calculation	Your calculator's answer is:	The CORRECT answer using significant figures is
3.24 m x 7.0 m	22.68 m^2	23 m^2
100.0 g ÷ 23.7 cm^3	4.219409283 g/cm^3	4.22 g/cm^3
0.02 cm x 2.371 cm	0.04742 cm^2	0.05 cm^2
712 m ÷ 3.0 sec	237.333333 m/sec	240 m/sec
1818.2 lb x 3.23 ft	5872.786 lb•ft	5870 lb•ft
1.030 g ÷ 2.87 ml	0.35888 g/ml	0.358 g/ml

Practice 4:

Calculation	Your calculator's answer is:	The CORRECT answer using significant figures is
512 g x 2.1 l	g/l	g/l
35.7 m ÷ 61 sec	m/sec	m/sec
2.410 g ÷ 0.37 ml	g/ml	g/ml
9 µm ÷ 0.8 min	µm/min	µm/min
77.33 km ÷ 0.34 hr	km/hr	km/hr

4. CHEMISTRY IN THE BIOSCIENCE LABORATORY

A. *Elements are arranged into the Periodic Table*

The elements evolved from the event known as the Big Bang. The periodic table is a convenient way of organizing the known elements into groups (colored boxes below) with similar characteristics based on their electron arrangement. Most elements are well-behaved. They are organized into rows (periods) that represent increasing number of protons (and electrons). All atoms <u>except hydrogen</u> have neutrons. Hydrogen atoms consist of only one proton and one electron.

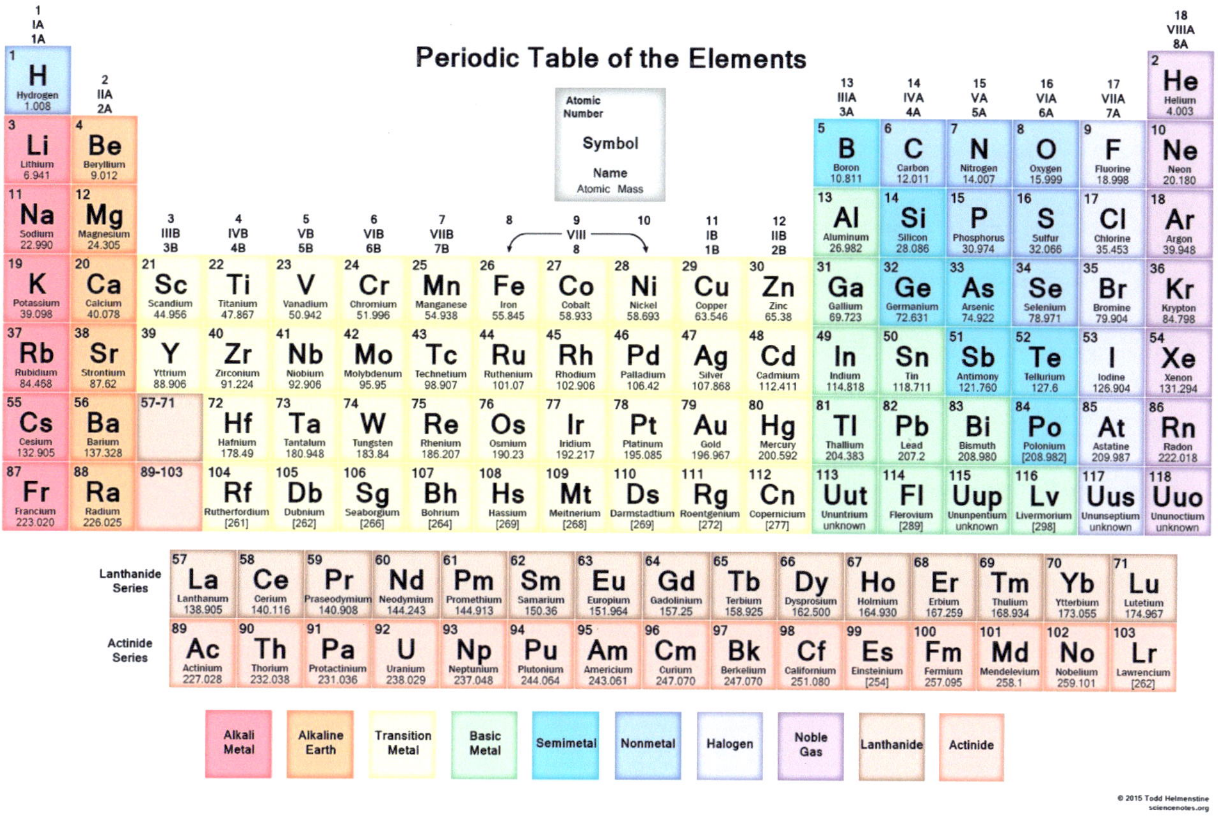

B. *Elements have mass*

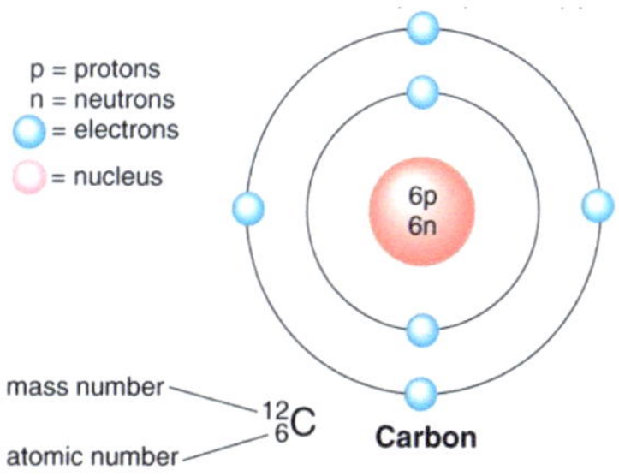

The mass of an element is its weight in the absence of gravity. When referring to atoms and molecules, <u>mass</u> (or molecular mass) is the appropriate term since gravity is negligible .

The mass of an atom is the sum of its number of protons and neutrons. The mass of an electron is negligible in comparison with the mass of protons and neutrons, so it does not get included in the total mass.

Protons and neutrons have the same mass. The mass of a single proton or neutron is 1.660539×10^{-24} g, which is one atomic mass unit (amu or u). The number 1.660539×10^{-24} g (1 amu) is 1/12th the mass of a C_{12} atom.

To standardize all elements to carbon, the molecular mass of each element is divided by 1 amu (1.660539×10^{-24} g) to give the ***relative*** molecular mass (abbreviated as M_r) or molecular weight (MW) M_r and MW have no units.

atom	atomic number	# of protons	# of neutrons	mass number
carbon	6	6	6	6+6 = 12 (actual 12.011)
oxygen	8	8	8	8+8 = 16 (actual 15.999)

The <u>actual</u> mass number for an atom will deviate from the perfect sum because of the occurrence of natural isotopes for each atom.

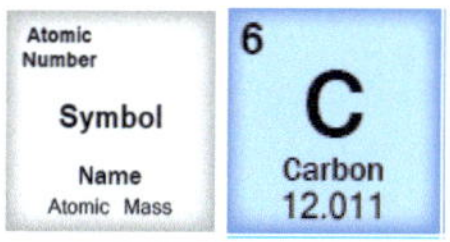

The mass of carbon is 12.011

Carbon is an important and versatile atom that is present in <u>most</u> biological compounds. This is due to the fact that carbon has many oxidation states (from +4 to -4) and can form bonds with different atoms.

C. *Mass of a compound*

To calculate the mass of a compound you need to know its 'parts' – the chemical composition of the compound.
Add the atomic masses of all of the atoms to get the molecular mass of the compound.

EXAMPLE — CALCULATE MOLECULAR WEIGHT OF NACL

1. Look up the atomic mass of the individual atoms from the periodic table:
   The atomic mass of Na (sodium) is 22.990
   The atomic mass of Cl (chlorine) is 35.453
2. How many atoms of each are present in NaCl?
   One Na
   One Cl
3. Add the atomic masses. 22.990 + 35.453
4. The relative molecular mass (M_r or molecular weight) = 58.443 (simplify to 58.44)

The molecular mass of NaCl is the sum of Na and Cl.

EXAMPLE — CALCULATE MOLECULAR WEIGHT OF GLUCOSE

1. Look up the atomic mass of the individual atoms from the periodic table:
   The atomic mass of C (carbon) is 12.011 (we will simplify to 12)
   The atomic mass of H (hydrogen) is 1.008 (simplify to 1)
   The atomic mass of O (oxygen) is 15.99 (simplify to 16)
2. How many atoms of each are present in the glucose molecule?
   C = 6
   H = 12
   O = 6
3. What mass is contributed by each atom? (multiply the atomic mass x the number of atoms)
   C = 12 x 6 = 72
   H = 1 x 12 = 12
   O = 16 x 6 = 96
4. Add the three masses = 72 + 12+ 96 = 180
5. The relative molecular mass (M_r or molecular weight) of glucose = 180

The chemical formula for glucose is $C_6H_{12}O_6$.

D. *Definition of a mole (abbreviation = mol)*

Scientists determined empirically that 1 mole (mol) of an element or a compound contains contains 6.022×10^{23} atoms or molecules, respectively. This number is called **Avogadro's constant (or Avogadro's number).**
One mole of any substance always contains 6.022×10^{23} particles (atoms or molecules), just as one dozen always means 12 regardless of the material being examined (a *dozen* eggs, a *dozen* loaves of bread, a *dozen* oranges, *etc.*).

base term	quantity for ONE	quantity for one-half	quantity for two
dozen	12	6	24
mole	6.022×10^{23}	3.011×10^{23}	12.044×10^{23}

Substitute MW for mole

6.022×10^{23} particles is an AWKWARD number to work with but you can use the molecular weight (MW) as a substitute.

1 mole of		molecules/mole	mass of 6.022×10^{23} molecules	one mole contains
sodium chloride (NaCl)	has	6.022×10^{23}	58.4 g	58.44 g
glucose ($C_6H_{12}O_6$)		6.022×10^{23}	180 g	180 g

Notice that a mole is independent of the mass of the compound. The **mass** of 1 mole of NaCl (58.44 g/mol) and 1 mole of glucose (180 g/mol) is quite different.

Vary the number of moles

If you have 24 eggs, you have 2 dozen eggs.
If you have more than (the symbol for more than is **>**) 180 g of glucose, you have >1 mole.

Example, 360 g of glucose equals two moles since 360 g ÷ 180 g/mol = 2.

Continuing the analogy, if you have 6 eggs, you have half a dozen or 0.5 dozen eggs.
If you have less than (the symbol for less than is **<**) 180 g of glucose, you have < 1 mole of glucose.

Example, $90\ g \div 180\ \frac{g}{mol} = 0.5\ mol$ thus 90 g of glucose equals 0.5 moles.

base term	quantity for one	quantity for one-half	quantity for two
dozen	12	6	24
mole of glucose	180 grams	180 ÷ 2 = 90 g	180 x 2 = 360 g

E. *Molarity - a measure of concentration (mol/l = M)*

Molarity (molar solution or **M**) refers to concentration in number of molecules per liter.

A 1 molar (1 M, 1 mol/l) solution of a compound is equivalent to 6.022 x 10^{23} molecules/liter total volume. Brackets [ ] are sometimes used to signify concentration.

1 mole/1 liter = 1 mol/l = 1 M

To make a **solution** of a compound and express the concentration in molarity, you have to know the molecular mass (*aka* molecular weight) of the compound (see above).

EXAMPLE— MAKE A 1 M SOLUTION OF SODIUM CHLORIDE (NACL)

The mass of 1 mole of NaCl = 58.44 g

Another way of saying this is that NaCl has 58.44 g/mol

Detailed calculations

Substitute 58.44 g in the equation for 1 M = 1 mol/l as follows

$$\frac{1\ mol}{liter}\ x\ \frac{58.44\ g}{mol} = \frac{58.44\ g}{liter}$$ **1 M NaCl = 58.44 g/l**

If you dissolve 58.44 g of NaCl in one liter of water, you will have a 1 M (molar) solution of NaCl (1 M NaCl).

Similarly, a **1 M** (molar) solution of **KCl** is **74.54 g/l** because 1 mol of KCl weighs 74.54 g.

EXAMPLE— WHAT IS THE MOLARITY (MOLES PER LITER) OF GLUCOSE IN COCA COLA?

For the sake of this example, assume that all of the sugar in *Coca Cola* is glucose. The molecular weight (M_r or MW) of glucose can be calculated from the formula for glucose – $C_6H_{12}O_6$ = 180 g /mol (see notes above for details). Hence 1 mole of glucose = 180 g glucose

1 liter of *Coca Cola* contains 52 g of sugar.

The math is straightforward. Set up an equivalence ratio — 180 g is to 1 mole as 52 g is to X (your unknown). Then solve for X.

$$\frac{180\ g}{1\ mol} = \frac{52\ g}{X\ mol} \rightarrow \underline{\text{Solve for } X}\ \rightarrow X\ mol = \frac{52\ g}{180\ g} = 0.29\ mol$$

The grams (g) cancel out and the mol in the *denominator moves to the numerator* and you are left with the number (#) of **moles.**

Answer: The initial volume was 1 liter, so the molarity of glucose in *Coca-Cola* = <u>0.29 mol per 1 liter or **0.29 M**</u>

How many <u>molecules</u> of glucose are in the 1-liter bottle of Coca Cola?

The 1-liter bottle of Coca Cola has 0.29 moles of glucose. Since you know that 1 mol = 6.022 x 10^{23}, the math is not very complicated

0.29 mol = 6.022 x 1023 x 0.29 = 1.75 x 10^{23} molecules.

Answer: If you consume 1 liter of Coca Cola you are consuming 1.75 x 10^{23} molecules of glucose.

EXAMPLE— CONVERTING FROM AN IMPERIAL (US) MEASURE OF VOLUME TO A METRIC MEASURE

How many moles of glucose are there in 12 ounces (oz) of Coca Cola?

Above, you calculated the molarity of *Coca Cola* as 0.29 M [*i.e.,* 0.29 mol/l). Now you are asked to calculate moles in a different <u>volume</u> of *Coca Cola*. The **concentration** of sugar, 0.29 M, stays the same, but you have 12 oz rather than 1 liter. You can convert 0.29 mol/l to the US measure or you can convert 12 oz to liter. This example shows how to convert ounces to liters in two steps.

1. Convert 12 oz to liters: (you can look up conversion tables on-line)

   12 oz = 12 oz ÷ 128 oz/gal = 0.094 gal

   1 gal = 3.78 liters

   0.094 gal x 3.78 liter/gal = **0.355 liter** (converted volume)

2. Now you can calculate how many moles of glucose are in 0.355 liters of a 0.29 M (mol/l) solution of *Coca Cola*:

Answer: 0.29 mol/l x 0.355 l = 0.103 mole or about 0.1 mole of glucose in 0.355 l (355 ml or 12 oz) of Coca Cola

5. MEASUREMENTS: PERCENT, PPM/PPB, AND MOLAR SOLUTIONS

Scientists work with solutions whose concentrations may be represented as A. percent solutions, B. parts per million (ppm) or parts per billion (ppb), or C. molar solutions.

Concentration (def) is the amount of solute (the ingredients or constituents that containing the solute plus the solvent).

EQUATION = $\frac{solute\ (the\ material\ being\ dissolved)}{total\ volume\ of\ the\ solution\ (the\ solute+solvent)}$.

The most frequently used solvent in biology is water. Solutions that are water based are referred to as ***aqueous*** solutions (from Latin: *aqua* meaning water).

- Solutes that have completely dissolved in the solvent are evenly distributed in the solution. Thus, every aliquot of the solution has the same concentration of solute per volume as the whole solution.
- The concentration of the solute is based on **mass (grams)** per total volume or **number of particles** (moles, mol) per total volume (liter, l) of solution.
- When diluting solutions, a known amount of solvent (such as water) is added to a known amount of solution.
- When concentrating solutions, some solvent is removed.

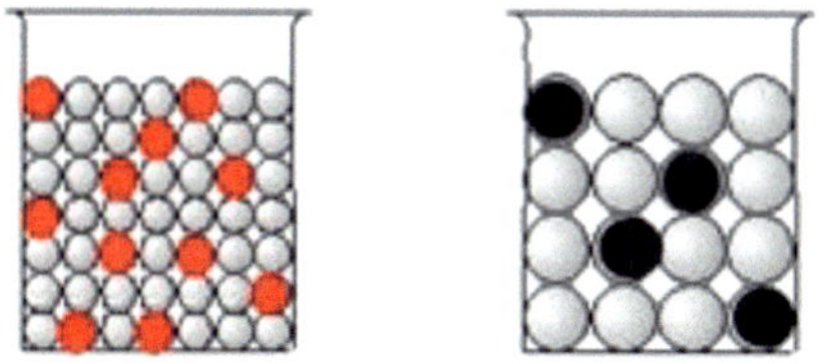

Fig 1. Examine the concentration of colored beads in these two mixtures. The concentration of red beads in the beaker on the left is 11 red beads/49 total beads= 0.22. To convert to percent, multiply by 100 to get 22%. The concentration of black beads in the second beaker is 4 beads/16 total beads x 100 = 25%.

A. *Percent solutions*

A percent solution is an example of concentration based on the **mass (or weight)/volume (W/V)** ratio.

A percent solution is always expressed as grams of solvent to a final volume of 100 ml.

$$\% = \frac{\text{grams (g) of solute}}{\text{100 ml of solution}}$$

How to make a 5% NaCl (sodium chloride) aqueous solution:

- weigh out 5 g NaCl
- dissolve in sufficient water to give a total volume of 100 ml.

This results in a NaCl concentration of 5 g NaCl/100 ml

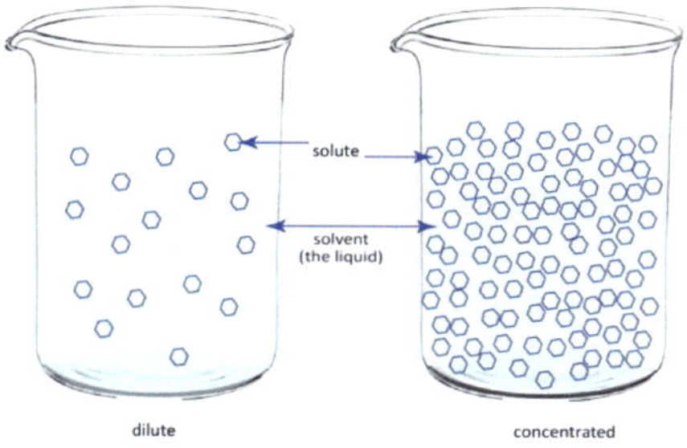

Fig 2. A concentrated solution contains large numbers of dissolved solute molecules; a dilute solution contains a lower number of dissolved solute molecules.

Sample question: How many grams of NaCl are in 400 ml (0.4 l) of a 5% NaCl (w/v) solution? Answer= 20 g

By definition, 5% (w/v) = $\frac{5\ g}{100\ ml}$

for 400 ml, multiple $\frac{5\ g}{100\ ml}$ x 4 = $\frac{20\ g}{400\ ml}$

Procedure: weight out 20 g of NaCl and add water to bring the total volume to 400 ml

Practice 1:
How many grams (g) of sugar are needed for 200 ml of a 10% sugar solution?

Practice 2:
You are preparing 250 ml of bacterial growth medium. The recipe states that you should add 1.5% agar. How much agar should you weigh out for 250 ml of medium?

Practice 3:
You have a stock of the herbicide Round*up* containing 10% glyphosate. How much (what volume) of this stock should you add to prepare 2 liters (total volume) of 2% glyphosate?

B. *ppm and ppb solutions*

The terms ppm and ppb are used when dealing with very low concentrations and refer the amount of **X** (such as a chemical) in water or soil. These terms are dimensionless quantities.

> **Remember:**
> **1 ml of water weighs 1 gram.**

PPM = PARTS PER MILLION (PARTS PER 1,000,000, 10^6)

Parts per million (ppm) $= \frac{\text{parts of solute}}{\text{one million parts of the solution}}$

1 ppm is equivalent to 1 milligram of a chemical per liter of water (mg/l) or 1 milligram of something per kilogram soil (mg/kg)

How to remember this: It is possible to determine ppm because we know that 1 ml of water = 1 g; this allows you to substitute weigh for volume in situations where water is the solvent (the norm for biological applications that we will deal with here).

If 1 ml of water = 1 g, then 1 ml of water = 1000 mg (since 1 g = 1000 mg)

1 liter of water weighs 1 kg or 1,000 g or 1,000,000 mg (notice that this is one million mg)

EXAMPLE

If you dissolve 3 mg of compound X in 1 liter of water, you can rephrase this as 3 mg in 1,000,000 mg of water → or 3 ppm

3 mg/L = **3 mg/1,000,000 mg** of water or three parts in one million.

PPB = PARTS PER BILLION (PARTS PER 1,000,000,000 OR 10^9)

Parts per billion (ppb) $= \frac{\text{\# of parts of solute}}{\text{one billion parts of the solution}}$

Equivalent ways to express ppm and ppb

1 ppm of compound X in a million parts (parts of water, solvent, soil, *etc.*)	**1 ppb of compound X in a billion parts (parts of water, solvent, soil, *etc.*)**
= 1000 µg/1,000,000,000 µg	= 1 µg/1,000,000,000 µg
= 1 mg/1,000,000 mg water	= 1 µg/1,000,000 mg
= 1 mg/1,000 g	= 1 µg/1,000 g
= 1 mg/kg	= 1 µg/kg
= 1 mg/l	= 1 µg/l

EXAMPLE

How would you express 1 ppm as a % solution?

1 ppm = 1 mg/10^6 mg

to convert to percent, multiply by 100

$\frac{1}{1{,}000{,}000}$ x 100 = $\frac{1}{10{,}000}$ % → this can be written as $\frac{1}{10^4}$ or 10^{-4} % which can be written as **0.0001%**

> **Practice 4:**
> How many total mg of nitrate can be present in 5 l of drinking water if the maximum allowable limit is 9.8 ppm?

C. *Molar solutions or molarity (M)*

Molar solutions are used in virtually all environmental, research, and medical labs.

Steps to calculate the mass (in g) needed to prepare a molar solution (M)

1.	Determine the molecular weight (grams/mol) from the periodic table (look in the Methods Manual)	$MgCl_2$ = Mg = 24.305 g/mol Cl = 35.452 X 2 = 70.904 g/mol Total MW is 24.305 + 70.904 = 95.209 g/mol
2.	A 1 M solution contains 1 mole in 1 liter (total volume) of solution (1 M = 1 mol/l)	95.209 g/1 mol x 1 mol/1 liter= 95.209 g/ l

> **THE RELATIONSHIP BETWEEN MW AND MOLARITY**
> *(abbreviations used: g = gram, mol = mole, l or L = liter)*
>
> **MW (molecular weight) is derived from the periodic table** = $\frac{grams}{mole}$
>
> M = $\frac{\text{moles of a solute (mol)}}{\text{volume of a solution (l)}}$ = mol/liter **If you factor in the molecular weight, M becomes** $\frac{\text{moles of a solute (mol)}}{\text{volume of a solution (l)}} x \frac{grams}{mole}$ = **grams/liter**

EXAMPLE (BASIC)

How do you make 1 liter of a 1 M solution of glucose?

- Determine the molecular weight (MW)
  glucose= $C_6H_{12}O_6$ = [6 C x 12 g/mol] + [12 H x 1 g/mol] + [6 O x 16 g/mol] = 180 g/mol
- Determine molarity (M) = 1 M = 180 g/mol x 1 mol/1 liter = 180 g/liter
  Dissolve 180 g of glucose in water to give a total vol of 1 liter

> **Use a ratio:** $\frac{M1}{g/l} = \frac{M2}{g/l}$
> M1 = molarity of solution 1
> M2 = Molarity of solution 2
> *Make sure that the units match (i.e. do not use liter and ml or gram and milligram in the same ratio).*

EXAMPLE (BASIC +)

How do you make 1 liter of a 0.4 M solution of glucose?

- Determine the MW (as above)
  glucose= $C_6H_{12}O_6$ (C)12 X 6 + (H) 12 X 1 + (O) 16 X 6 = 180 g/mol
- Multiply by the desired molarity
  If 1 M = 180 g/l, then 0.4 M = 0.4 x 1 mol/l = 0.4 x 180 g/l= 72 g/l
  Weigh out 72 g of glucose and dissolve in water to give a final (total) vol of 1 liter

Practice 5:
How do you make 100 ml of 0.5M NaCl solution? It helps to find the answer without skipping steps:

1st Calculate the MW
2nd Determine 1M (the g/l of 1 mol)
3rd Determine the g/l of 0.5 M
4th Determine g in 100 ml of 0.5M

EXAMPLE (ADVANCED)

How do you make 200 ml of a 0.4 M solution of glucose?

- First determine the MW of glucose
  - glucose= $C_6H_{12}O_6$ = [6 carbon x 12 g/mol] + [12 hydrogen x 1 g/mol] + [6 oxygen x 16 g/mol] = 180 g/mol
- If the MW of glucose is 180 g/mol, then *by definition*, 1M glucose contains 180 g/l
- Second, plug the numbers into the ratio: $\frac{g/l}{M1} = \frac{g/l}{M2}$ **or** $\frac{180\ g/l}{1\ M} = \frac{??\ g/l}{0.4\ M}$ to determine how much glucose is needed for a 0.4 M solution.
- Multiply both sides by 0.4 M : $\frac{180\ g/l}{1\ \cancel{M}}\ x\ 0.4\cancel{M} = \frac{??\ g/l}{\cancel{0.4\ M}}\ x\ \cancel{0.4M}$ → 180 g/l x 0.4 = ?? g/l = 72 g/l
- Third, calculate the amount of glucose you need for 200 ml of 0.4M glucose
- If there are 72 g in 1 l of 0.4 M glucose, then there are X g in 0.2 l
  - ratio = $\frac{72\ g}{1\ l} = \frac{x\ g}{0.2\ l}$
  - solve for *x*: X = $\frac{72\ g}{1\ l}$ x 0.2 l = 14.4 g
- Dissolve 14.4 g of glucose in water to a final volume of 200 ml

Practice 6:
How do you prepare 500 ml of 1M potassium phosphate (KH_2PO_4) buffer?

1st Determine the MW
KH_2PO_4 = _____ K + ______ H +_____ P + _____ O
MW of KH_2PO_4 = __________
2nd Determine the number of grams in 1 mol = ________ g/mol

3rd Determine the grams/liter for a 1 M solution = _________ g/liter

4th Determine the number of grams in 500 ml of a 1 M solutions = _________ g (per 0.5 l)

6. WORKING WITH CONCENTRATED (STOCK) SOLUTIONS

When you work in an environmental, research, and medical laboratory, you will occasionally have **stock solutions <u>or</u> microbial cell cultures** that require dilution prior to use because they are **too concentrated**. Some examples of stock chemical solutions are antibiotics, vaccines, chemicals, dyes, buffers etc.; microbial cell cultures contain many millions of individual cells in a small volume (e.g. 10^8 or 10^9 cells in one milliliter). Concentrated stock solutions such as a 1M or 10% solutions of NaCl or glucose (described in other sections) save space, prolong shelf-life (some chemicals are more stable in concentrated solutions) and increase reproducibility. You will save time if you have a concentrated stock solution that you can use to quickly dilute into a working stock.

A. *Concentrated stock solutions*

We dilute a concentrated solution to prepare a working solution. A precise dilution occurs when you mix a known amount of solvent, such as water, with a known amount of a concentrated solution (X). The act of diluting will decrease the concentration of X.

Fig 1. Mixing 1 container of frozen concentrate with three containers of water will yield 1X (normal) juice (and less plastic waste).

Example of a simple dilution

The food court uses frozen blueberry-pomegranate juice concentrate to make a 1X working solution (drinkable juice). The instructions on the container read: add one (1) container of frozen concentrate to three (3) containers of water. This results in four (4) volumes of drinkable blueberry-pomegranate juice. The dilution FACTOR is 1 to 4, also written as $\frac{1}{4}$

How to calculate the dilution factor

THE FORMULA IS : $\frac{\textit{volume of concentrate}}{\textit{total volume (concentrate+water)}} = \frac{1}{1+3} = \frac{1}{4}$

4-4-4 4 parts, 4X, $\frac{1}{4}$

The four (4) in the $\frac{1}{4}$ ratio tells us that concentrated juice is a **4X concentrate** solution and requires a 1:4 (or 4-fold) dilution to yield drinkable blueberry-pomegranate juice.

You can calculate the concentration or volume using the equation: **C1×V1 = C2×V2**

- C1 is the concentration (such as percent or molarity (moles/L)) of the initial, concentrated solution (stock solution)
- V1 is the volume of the concentrated solution of the concentration C1
- C2 is the concentration of the dilute solution (after more solvent has been added)
- V2 is the volume of the dilute solution if the concentration C2

EXAMPLE

Make a dilution using a percent stock solution

How much of a 40% glucose (dextrose or D-glucose) stock solution do you need to add to 1 L (remember that this is the total volume) of YPD (yeast extract, potato, dextrose) broth to achieve a final concentration of 2%?

C1 = 40%
V1 = unknown
C2 = 2%
V2 = 1 L

Plug the numbers into the formula: 40% × V1= 2% × 1L

Solve for V1:
V1 = 2% x 1L ÷ 40% = 0.05 L (% units cancel)
V1 = 0.05 L Convert from L to ml
V1 = 50 ml

Answer: To make 1 L total, you will add 50 ml of 40% glucose to 950 ml of YPD broth

Practice 1:
You have concentrated Roundup that contains 50% glyphosate. You need to make 200 ml of 2.5% glyphosate. How much of the concentrated Roundup will you use?

EXAMPLE

Make a dilution of a molar stock solution

How many milliliters of a 12 M HCl (hydrochloric acid) stock do you need to prepare 50 ml of a 3 M HCl solution?

(M = mol/L). Here are the three known values: 12 M HCl, 3 M HCl and 50 ml.

Use the equation: C1×V1 = C2×V2

Your starting stock is 12 M HCl — this is C1

The desired final volume is 50 ml — this is V2

The desired final concentration is 3M — this is C2

Solve for V1

C1 ×V1 = C2 × V2

12 M x V1 (ml) = 3M x 50 ml

V1 = $\frac{150\ M \bullet ml}{12\ M}$ = 12.5 ml

To prepare 50 ml of 3M HCl, you need 12.5 ml of 12M HCl plus 37.5 ml H_2O

EXAMPLE

Advanced dilution question

A large animal veterinarian is called to a ranch to vaccinate a herd of cows against the disease brucellosis. The vaccine stock is supplied frozen as a 100X concentrate in 5 ml aliquots that are diluted with 0.85% saline (this is the diluent) immediately prior to use. Each cow will receive a shot of 0.5 ml dose of the 1X Brucellosis vaccine. How many cows can be vaccinated from 5 ml of the original 100X concentrate?

By definition, the concentration of a ready-to-use solution is 1X.
10X means that a solution is 10 times more concentrated than normal. 50X is 50 times more concentrated.

You can use math to your advantage. What number do you need to divide 50 by to get 1? $\frac{50x}{?}$ = 1

Answer $\frac{50x}{50}$ = 1

Answer:

A. Calculate the volume of working solution you will have upon dilution of the vaccine stock.

Here is what is known: 100X vaccine, 5 ml volume, 1X desired final working solution.

C1 × V1 = C2 × V2

100X × 5 ml = 1X × V2

solve for V2 = 500X • ml ÷ 1X = 500 ml

Important: 500 ml is the final *total* volume. Since V1 was 5 ml; the volume of saline to add is 495 ml

Here is another way to calculate the volume of working solution

First, consider the dilution: a dilution of 100X to 1X requires a 1:100 dilution. 1:100 = $\frac{1\ part}{(100-1=99)+1\ parts}$

Hence you need 1 part of the 100X vaccine stock + 99 parts of diluent (*i.e.,* water) to yield the 1X solution. The table below shows alternative ways to think about this dilution and arrive at the correct answer in row 4.

Stock	Vol of stock		diluent (saline)		= this volume of 1 X solution
1. 100X concentrate	1 part		99 parts		100 parts
2. 100X vaccine	1 ml	+	99 ml	=	100 ml
3. 100X vaccine	5 x 1 ml		5 x 99 ml		5 x 100 ml
4. 100X vaccine	5 ml		495 ml		500 ml

B. Once you have diluted the vaccine, calculate the number of animals to be treated.

Each animal receives a dose of 0.5 ml of the 1X vaccine. To calculate the number of animals that can be vaccinated with 500 ml of the 1X vaccine, the 500 ml must be divided by the dose/animal, so $\frac{500\ ml}{0.5\ ml/animal}$ = 1000 animals can be vaccinated.

Practice 2:

To prepare a 1X working solution from 1 ml (vol) of a 5X concentrated solution, you will need to add ____ ml water.

The total volume will be ______ ml.

Practice 3:

Prepare different volumes of 1X working solution from concentrated stock solutions using the numbers in the table below. The first three have been done for you.

Practice 4:
How many milliliters of a 2 M $CuSO_4$ solution must be added to water (total volume 160 ml) to achieve a 0.05 M $CuSO_4$ solution?

Practice 5:
Use the data provided to calculate answers in the blank spots.

Stock solution	Dilution	Parts of stock	Parts of water	Total # parts	Final volume	Actual volume of stock solution	Water = (total vol – vol of stock)
3X	1:3	1	2	3	10 ml	10 ml/**3** = 3.3 ml	6.7 ml
5X	1:5	1	4	5	10 ml	10 ml/**5** = 2 ml	8 ml
5X	1:5	1	4	5	200 ml	200 ml/**5** = 40 ml	160 ml
2X		1			500 ml		
20X		1			60 ml		
10X		1			500 ml		
6X		1			12 µl		
25X		1			1 l		

B. *Solutions based on molarity [unit: mol/liter or M]*

A 1 M (1 molar) solution contains 1 mole of a solute dissolved in a total of 1 liter of solution (in biology water is the most often the solvent). The same equation **C1×V1 = C2×V2** that was in the primer last week (example B) applies when diluting concentrated molar solutions.

- C1 is the concentration (such as percent or molarity (moles/L)) of the initial, concentrated solution (stock solution)
- V1 is the volume of the concentrated solution of the concentration C1
- C2 is the concentration of the dilute solution (after more solvent has been added)
- V2 is the volume of the dilute solution if the concentration C2

EXAMPLE

What volume of 1 M glucose do you need to M9 base to produce 1 L (total) of M9 medium + 20 mM glucose? Consider what is known:

V1 = ?
C1 = 1 M glucose
V2 = 1 l (liter total volume)
C2 = 20 mM (the desired final concentration)

Use C1×V1 = C2×V2 equation to calculate dilutions

Important → ***V1/V2 must have matching units***
C1/C2 must have matching units

In Example 1, C1 is M but C2 is mM. Therefore, you must convert one of these to match the other.
We will convert 20 mM to M. 20 mM = 0.02 M
Next, plug the numbers into the formula: 1M × V1 = 0.02 M × 1 L
Solve for V1: V1 = 0.02 M x 1 L / 1 M The units, **M,** cancel
V1 = 0.02 L Convert to ml
V1 = 20 ml

Answer: I need to add **20 ml** of 1M glucose to M9 medium. Note that the total volume will be 1 liter, so you will add 20 ml + 980 ml of M9 medium (1 liter of M9 – 20 ml = 980 ml).

EXAMPLE

TE 10:1 is a buffer commonly used to dissolve DNA. TE 10:1 is shorthand for 10 mM Tris, 1 mM EDTA. The stock solutions are 1 M Tris and 0.5 M EDTA.

How much of each of these stock solutions is required to make 100 ml TE 10:1 buffer?

Calculations for Tris

C1 = 1 M Tris
V1 = unknown
C2 = 10 mM Tris
V2 = 100 ml = 0.1 L

Convert 1 M Tris to 1000 mM so that units match

C1×V1 = C2×V2 →	1 M × V1 = 10 mM × 0.1 L	
	V1 = (10 mM × 0.1 L) /1000 mM	The units, **mM,** cancel
	V1 = 0.001 L	Convert to ml
	V1 = 1 ml	

Answer: You need 1 ml of 1 M Tris per 100 ml to get a final concentration of 10 mM Tris.

Calculations for EDTA

C1 = 0.5 M EDTA
V1 = unknown
C2 = 1 mM EDTA
V2 = 100 ml = 0.1 L

Convert C1 to 500 mM so that units match.

C1×V1 = C2×V2 →	500 mM × V1 = 1 mM × 0.1 L	
	= (1 mM × 0.1 L) / 500 mM	The **mM** units cancel
	V1 = 0.0002 L	Convert to ml
	V1 = 0.0002 L = 0.2 ml	Convert to µl
	V1 = 0.2 ml = 200 µl	

Answer: You need 200 µl of 0.5 M EDTA stock solution per 100 ml final solution to get a 1 mM concentration of EDTA.

SUMMARY: 1 ml of 1 M Tris + 200 µl 0.5 M EDTA + 98.8 ml water = 100 ml TE 10:1

Practice 6:

You add 20 ml of a 3 M NaCl to 180 ml H_2O to make solution A. Calculate the concentration of NaCl in solution A.

Practice 7:

You need ____ ml of a 50X surfactant stock solution to make 4 liters of a 1X working solution.

Practice 8:

You need to get rid of fungus growing on your grapes. The preferred treatment is 10 mM $CuSO_4$ (copper sulfate). You need to make as much 10 mM $CuSO_4$ as possible, but you only have 50 ml of a 0.5 M $CuSO_4$ stock. If you use all 50 ml of the 0.5 M stock, how much water should you add to the 50 ml to create a 10 mM $CuSO_4$ solution?

7. WORKING WITH LIVING PLANTS AND ORGANISMS

A. *Growing plants for science experiments*

Plants are excellent models for scientific work. Careful cultivation is critical for any plant-based experiment. Seeds need warm temperatures and regular water to germinate in soil. To sustain growth, seedlings need air (particularly, carbon dioxide), warm temperatures, nitrogen, trace elements, and regular water to grow into plants.

Preparing soil for seeding

1. Commercial potting soil has been heated to kill weed seeds, but it also has a negative impact on the beneficial microbial population in the soil.
2. Take the amount of soil that you will be needing for your pots (use about 1 C per 4-inch pot) and add it to a soaking container. As shown in Fig 1, add enough water to saturate the soil, mix well with your hands or a spoon, and let the soil rest for about 20 min. In the meantime, set up your pots and label them.
3. After soaking, transfer the soil to the pots — make sure there is a tray beneath the pots to capture the excess water. Fill the pots and lightly tap, but do not compress the soil too much.

Fig 1. To ensure that the soil is moistened, use your hands to mix water into the soil before you plant seeds. It can be difficult to moisten commercial potting soil because it is very dry and soil particles will float on top of the water without holding the water.

Seed planting tips

- Do not over crowd: four seeds per 4-inch pot gives individual plants room to grow. The exact number of seeds will depend on the size of the seeds that are planted.
- Pay attention to the type of seed – read information on-line about planting depth. Seeds planted too deep/shallow may not germinate properly; most seeds need to be covered by soil to germinate; some seeds must have light to germinate.
- Pay attention to **germination time** (number of days it takes a seed to sprout when placed in favorable conditions), **germination temperature,** and **germination rate** (the percentage of seeds that germinate; this number decreases as seeds age). Some information about germination temperature, time, and rate is shown in the table on the next page. More information is available on-line.
- Set up a watering schedule. *The plants must be checked at least every other day*. Keep track of the volume of water you are adding and when it is added.
- Grow your plants in one location; they should receive the same amount of light and similar temperature conditions.

Seed Germination Data

Seed type	Seed depth, cm	Germination temperature, °C	Average days to germination*
Soybeans	2-4	12-32°	7
Corn	3-5	10-20°	5-8
Alfalfa	<1	15-25°	2-5
Arugula	<1	22°	2-4
Mung bean	2-3	15°	2-3
Radish	1	7-10°	4-10

*Germination under optimum conditions of temperature and moisture

B. *Making broth and agar media to culture microorganisms and animals*

Using balances to weigh out ingredients: the top-loading balance

A top-loading balance is typically used when for gram amounts of chemical. These balances are accurate to one or two digits.

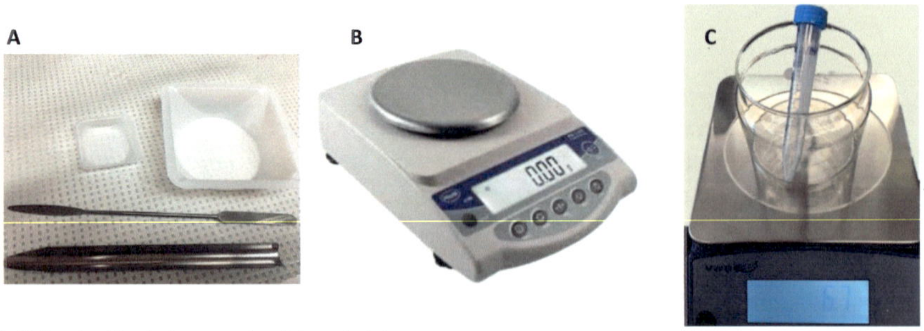

Fig 2. Top-loading balance and weigh materials
A. Weigh boat (or paper) and spatula must be used; B. standard balance; C. Beakers can be used to prop up vessels like centrifuge tubes.

Proper use of a top-loading scale

1. Obtain chemical, a weigh boat, and a spatula
2. Turn on scale
3. Place weigh boat on scale
4. Subtract the weight of the empty container (the weigh boat in this case) by pressing the tare button. The dial (read-out) should then display 0.00 g
5. Using the spatula, carefully add small amounts of your material to weigh boat until you reach the desired weight. Work close to the scale, do not drop from high up as this generates an aerosol.
6. Read the display
7. Pick up weigh boat and use it to funnel the material into a vessel such as a beaker
8. **Clean up weigh area**, wash spatula and weigh boat

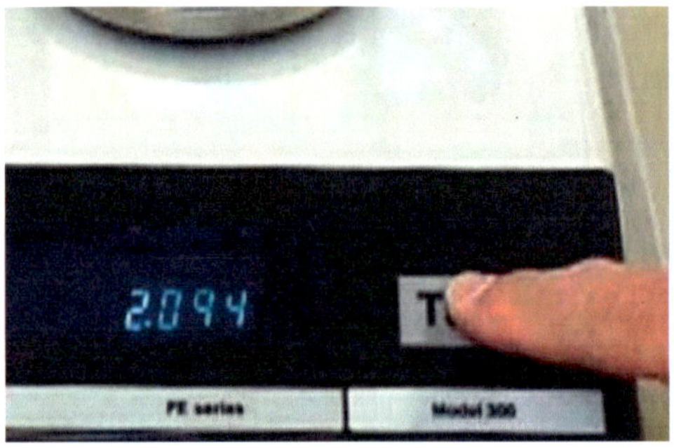

Practice 1:
The reading on this top-loading balance is ___________g.

The last digit on this top-loading balance reads _______mg.

The person's finger is pressing the ___________ button.

On this top-loading balance, could you accurately weigh a sample that is 0.0016 g?
Yes No

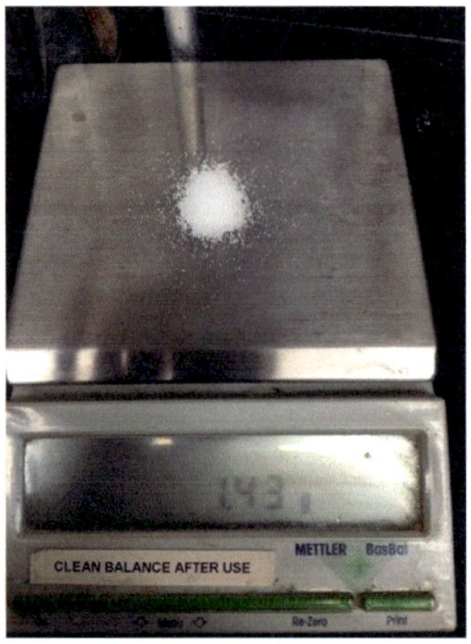

Practice 2:
What mistake has the operator made in this photo?

Using balances to weigh out small amounts: the analytical balance

Analytical balances are ***more sensitive*** than top loading balances. To prevent interference from **drafts and humidity** (in case of hydroscopic substances, water attracting) the weighing platform is in an enclosed chamber that can be accessed by a door. This door must be **closed** to ensure accurate measurements

To weigh out small amount directly into the desired small vessel (*i.e.* an Eppendorf tube), the Eppendorf tube can be used instead of weigh paper of a weigh boat, tared to zero the balance and the chemical is added directly to the Eppendorf tube.

Practice 3:

The readout on this analytical balance shows ________ g.

Write this out as milligrams: ____________________

Making liquid (broth) or agar medium to grow organisms

- Medium = singular; media = plural
- Living organisms need food. The media contain nutrients (source of carbon, nitrogen, trace minerals, vitamins *etc.*) that enable organisms (microbes, fruit flies) to grow.
- If you have AGAR in the medium, the final product will be solid medium, typically for pouring into Petri plates. Most often, the concentration of agar for solid medium is 1.5% (1.5 g/100 ml). Some media can be purchased as mixes that already contain the agar.
- Agar solidifies at about 40°C. After autoclaving, agar will solidify. After the medium has cooled to ≈50-55°C, the Petri plates can be poured (wear a glove).
- If the label says, BROTH, this will be a liquid medium. If you add agar to broth, it will be some a solid medium.
- All media must be autoclaved prior to use.

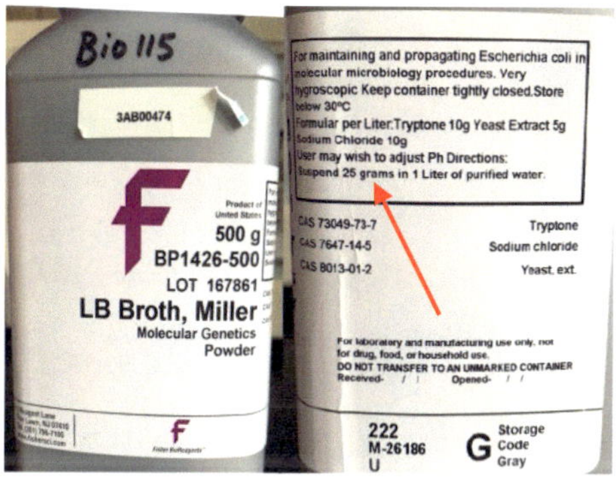

Fig. 3. The media bottle in this photo contains powder for a BROTH. *Broth is a liquid medium. if you plan to pour plates for <u>solid</u> medium (plates), make sure that you use medium that says AGAR.*

Preparing broth medium

1. Calculate the amount of media that needs to be made. Each broth tube holds ≈3- 5 ml of broth. If you need 20 tubes each containing 5 ml, you should prepare 100 ml of broth.
2. Follow package instructions for preparation. Instructions are typically written for 1l (1000 ml) of media. In the figure here for Luria-Bertani (LB) broth→, the ingredients tryptone (10 g), yeast extract (5 g), and NaCl (10 g) have been combined for you. Weigh out 25 g for 1 liter of this medium. Note that this formulation does not contain agar. If you want to make LB agar for plates, you must add agar.
3. If you need *less* than 1 liter, calculate the amount needed. For example, if the instructions state 25 g for 1 l and 100 mL is desired, use a ratio to calculate the amount needed:

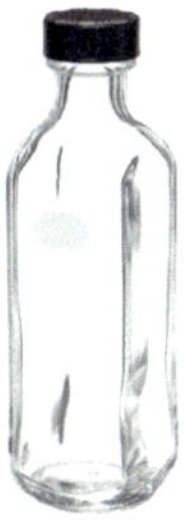

Fig 4. This is a "milk" bottle. Agar medium, such as NA (Nutrient Agar) or LB (Luria-Bertani agar) can be made in milk bottles in 100 ml aliquots, sterilized and stored for several months. Storing agar medium in bottles prevents loss of water from the medium and if condensation does occur it will be reintroduced into the medium when melted by heating.

4. Prepare your medium in a bottle large enough to hold the liquid and leaves some air space (*i.e.* prepare 100 ml of medium in a bottle — the bottle must be large enough to hold 100 ml plus room for expansion of water during autoclaving).
5. Open the culture medium container away from draughts and moisture. Avoid inhaling the powder and prolonged skin contact. Weigh the powder quickly, accurately and without creating 'clouds of dust'. Reclose the container as soon as possible.
6. Pour half the required volume of distilled water (in this example, 50 ml) in the vessel, then add the weighed powder and swirl to dissolve. Pour the rest of the distilled water (50 ml) down the sides of the vessel to wash any adherent medium back into solution. This is important because dry powder above the level of the water may not be sterilized in the autoclave and may be a source of contamination.
7. In general, you do not need to adjust the pH of dehydrated medium; it has been adjusted by the manufacturer.
8. When the powder has dissolved (prior to autoclaving) you can dispense the dissolved broth into tubes if desired. Place loose caps on filled tubes. Alternatively, you can autoclave the broth in the bottle.

9. Label the tubes/bottle with autoclave tape and a tape label with the following:
    a. media type (*i.e.* LB broth)
    b. the date
    c. section- group #, your initials

LB Broth 8-19-21 MR

10. Autoclave for 20 min. After sterilization is complete, remove the test tube rack, and allow the broth (in the milk bottle or in tubes if you chose to dispense the liquid) to cool at room temperature.

Preparing medium containing agar

1. Calculate the amount of medium that you need.
2. Each Petri plate requires ≈25 – 30 ml of agar. If 10 plates are needed, ≈250 ml of agar is needed.
3. Follow package instructions (or recipe) for preparation. Instructions are typically written for 1 l (1000 ml) of medium. If less is desired calculate the amount needed as shown in the box below.
4. Prepare your medium in a bottle large enough to accommodate the final volume you desire; you want some empty space. For example, prepare 400 ml of media in a 500 ml bottle (Fig 5).
5. Label the tubes/bottle with autoclave tape and a tape label with the following:
    a. media type (*i.e.* LB Agar)
    b. the date
    c. your initials

LB agar 8-18-21 KH

6. Add powder to bottle first, and then fill with necessary amount of water
7. Cap the bottle and swirl to mix the medium. [The agar will not dissolve until the solution is heated to boiling].
8. *Optional:* check pH if necessary. Most commercially available powder mixes <u>do not</u> need adjustment.
9. Autoclave on a <u>liquid</u> cycle (this takes about 1 hr to heat and decompress).
10. Once sterilization is complete, the medium will be allowed to cool.

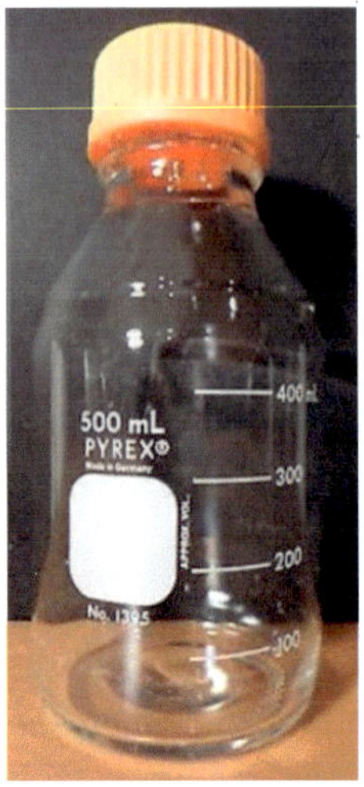

Fig. 5. This Pyrex glass bottle is a good choice for autoclaving medium.

If there are 25 g/L (or 25 g/1000 ml), how many grams (X) are there in 100 ml?

<u>This can be written as a user-friendly ratio</u> $\frac{25g}{1000\ ml} = \frac{X}{100\ ml}$

→ multiply both sides by 100 ml $\frac{25g}{1000\ ml} \text{ x } 100 \text{ ml} = \frac{X}{100\ ml} \text{ x } 100 \text{ ml}$

→ simplify $\frac{25g}{100\cancel{0\ ml}} \text{ x } \cancel{100 \text{ ml}} = \frac{X}{\cancel{100\ ml}} \text{ x } \cancel{100 \text{ ml}}$

→ $\frac{25g}{10} = \text{X}$ X = **2.5 g**

Pouring agar into Petri plates

1. You will need to melt the agar in the microwave before pouring into Petri plates.
2. Check cap and loosen slightly.
3. Place bottle into a beaker with water.
4. Microwave at **50% power** until melted; initially microwave for 2-3 min, then remove the bottle.
5. Put on rubber gloves, retighten the cap and gently invert the bottle – you are looking to chucks of agar that need to be dispersed. Chunks of agar are common because the contents of the bottle do not heat evenly. **If agar is not mixed evenly, you have chunks of agar in your plate. Some plates may not solidify while others will be too hard.**
6. Microwave for another 2-3 min if necessary.
7. When the agar is all liquified, place the hot agar in a 50°C water bath to equilibrate; ≈ 20 min
8. Pour plates: pour approximately 25 mL of agar into the plates. The bottom of the plate needs to be covered. If necessary, swirl the plate slightly in order to evenly disperse agar.
9. When agar has solidified in the Petri plates, turn them upside down overnight to dry to reduce condensation.
10. To keep plates from desiccating, you can store the plates in plastic sleeves or in the plastic show boxes in the lab.
11. Plates can be stored in fridge for a few weeks.
12. Be sure your plates are marker so that you know the type of medium (*i.e.,* use a brown Sharpie to label LB agar, green for R2A agar, *etc.*).

C. *Preparing solutions*

Scientists routinely need solutions, such as buffers, diluents, assay solutions and stabilizing solutions. In life sciences, these materials are often aqueous, meaning water-based. Some solutions are prepared as concentrated stock solutions that are stored in the laboratory and diluted to the working concentration when needed.

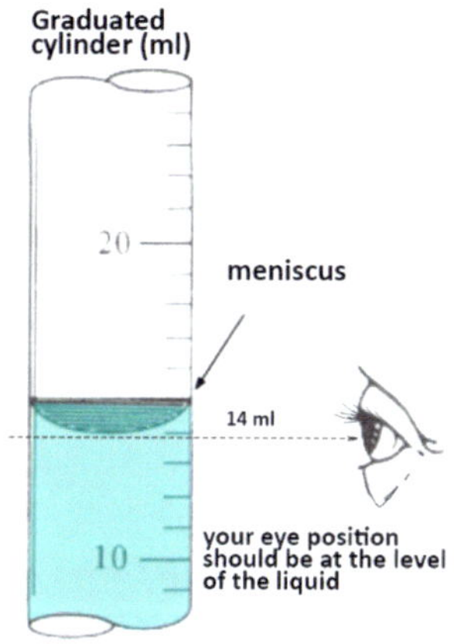

Fig 6. Read the bottom of the meniscus for an accurate reading of the volume, here 14 ml.

Use the following steps to prepare a solution of a chemical

1. Determine the volume of solution needed.
2. Calculate the amount of chemical required for the desired volume
3. Take these items to the balance:
   a. the chemical
   b. an Erlenmeyer flask or beaker
   c. a graduated cylinder
   d. weigh paper or boat
   e. a spatula
4. Weigh out the chemical on weigh paper/boat. Remove small amounts of chemical using the spatula; do not pour from the bottle. Be sure to follow the weighing instructions at the beginning of the chapter. Clean up any spilled compound.
5. Add some of your solvent (such as water) to the flask/beaker. Carefully and slowly transfer the chemical to the Erlenmeyer or beaker. Swirl gently to dissolve the chemical as you add it to the solvent.
6. When solute is dissolved, transfer the solution to the measuring cylinder and adjust to desired volume with water. Look at the bottom of the meniscus to determine when you have reached the correct volume (see Fig 6).
7. Transfer the solution to the desire storage container (bottle, Erlenmeyer etc.)
8. Label the container with the type of solution, the concentration, the date, your initials, section and group #. First responders need proper labeling on all laboratory chemicals.
9. The solution can be sterilized either by autoclaving, if the chemical is heat stable or by sterile filtration (see aseptic technique), if the chemical is labile (sensitive to heat).

D. *Troubleshooting*

Problem	Possible reason	Possible solution
Agar plates are not solidifying	Medium was not mixed well when re-melted in the bottle	Make sure that you mix stock agar well after melting before you pour new plates
	pH maybe too low, agar does not set below pH4	Do not adjust agar medium to low (pH 4) or high (pH 10) values
	Plates were poured from melted medium after autoclaving, but the agar had settled to the bottom of the bottle; the liquid at the top had insufficient agar to solidify in the plate.	Make sure that you mix medium well after melting in the microwave before pouring plates
My liquid (broth) medium is solid	You didn't read the label carefully and used, for example, LB agar instead of LB broth	READ the labels
My agar medium is liquid	You forgot to add agar to the medium	Agar media contain 1.5% agar; note that some premixed media already contain agar. Be sure to read the labels
Agar is lumpy	The lumps are from chunks of agar that were solid when you poured the medium; the agar medium was not completely melted and mixed enough	Make sure that you mix medium well after melting in the microwave before pouring plates
Surface of the agar is wet	Condensation on the medium	Agar was too hot when poured; cool medium to about 50°C (warm to the touch) before pouring The plates were not allowed to dry properly before refrigeration Plates were not inverted during incubation
Medium discolors after autoclaving	Some darkening is normal but drastic changes are not normal.	Re-examine the recipe and make sure that you did not add ingredients that should have been added <u>after</u> autoclaving
Nothing grew on my agar plates	You used the wrong medium You inoculating loop was too hot The incubation temperature was too high The incubation time was too long/short	Make sure that you use the correct medium, *e.g.*, yeast likes PDA or YPD, *E. coli* grows well on LB or NA

8. MICROPIPETTES AND PIPETTING

As a scientist, you need to use precision micropipettes and be able to dispense a specified volume of a solution accurately.

A. *Types of pipettes*

- **Glass pipettes** are used for manipulating volumes between ≈0.1 and 25 ml. Sterile glass pipettes are stored in metal cans. Use a pipet aide to draw and expel liquid. Your instructor will demonstrate how to use glass pipets to distribute nonsterile liquids (such as buffers) and sterile media.
- **Transfer pipettes** (either plastic or glass Pasteur pipettes): are used when precise measurements are not needed. Use with a rubber bulb.
- **Micropipettes** (shown on the right) are used for <u>small volumes</u>**.** These instruments are expensive. LEARN TO USE THEM PROPERLY

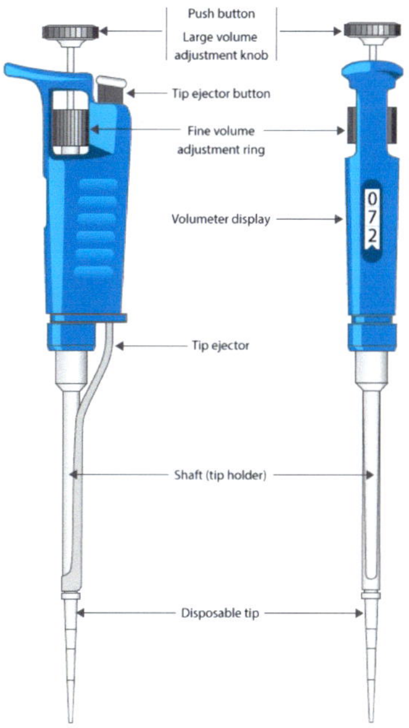

Fig 1. Parts of a micropipette

PROPER USE OF THE MICROPIPETTE

- You <u>must</u> use a **disposable plastic tip** with a micropipette. It attaches to the **plastic shaft**.
- The **plunger** is used to take up and dispense the volume. The plunger has <u>two stops</u>; the first stop is calibrated to take up the desired volume and the second stop is used to dispense that volume.
- After dispensing the liquid, the used tip is ejected using the **tip ejector** button.

B. *Micropipettes come in different sizes and use specific tips*

The most commonly used micropipettes in this lab are P20, P200, and P1000. The number on the plunger button refers to the **MAXIMUM** volume that can be dispensed with the device. *Learn the RANGE for each instrument.*

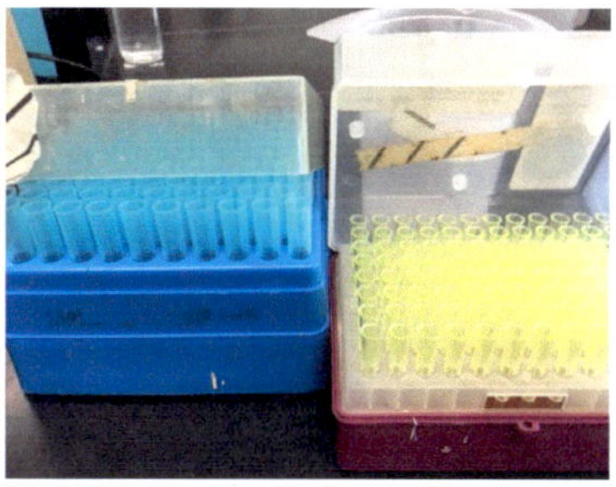
Fig 2. Tips are stored in boxes of 96; they are sterilized by autoclaving. The tip box should only be opened briefly to avoid contamination.

Micropipette settings

Type of pipettor	Range for this pipettor
P1000	100 or 200 to 1000 μl (may depend on the manufacturer)
P200	20 μl to 200 μl
P20	2 μl to 20 μl
P2	0.2 to 2 μl

To transfer a specific volume of liquid, you must set the **volume adjustment** dial reading the volume in the **volume indicator** window. Look at Fig 3.

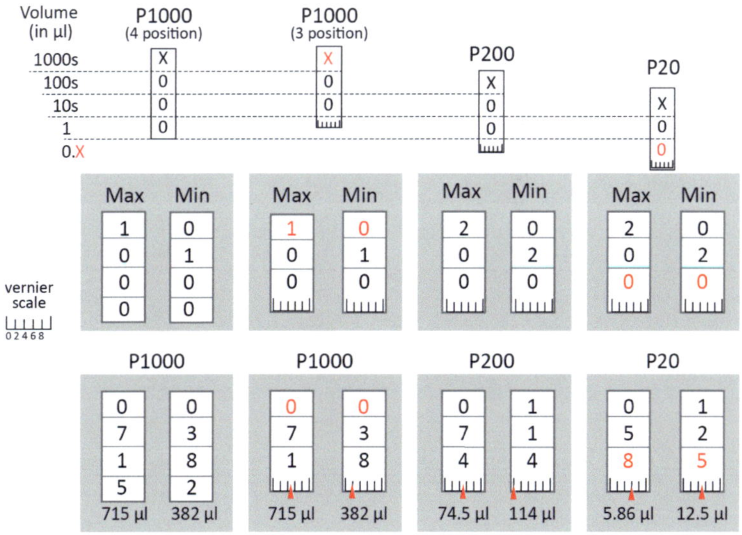

Fig 3 (left). This figure shows the relationship between volume and number position on four micropipette sizes (1000, 200, 20) commonly used in teaching labs.
Top row: alignment showing the position of the volume levels for different instruments
Middle row: examples of P1000, P200 and P20 micropipette settings, including the minimum and maximum
Bottom row: volume settings examples; note that the vernier scale is used to mark the single digit or 1/10th position for some micropipettes.

C. *Proper use of the P1000 micropipette*

The **P1000** micropipette is used in chemistry and biology laboratories. The instrument is used to aliquot between 0.2 ml (200 µl) and 1 ml (1,000 µl). Some models are accurate for volumes as low as 0.1 ml (= 100 µl), but none can aliquot volumes larger than 1 ml (1,000 µl). **Please do not exceed 1000 µl on the P1000; this will damage the instrument.**

Fig 4. This is a dial on a P1000. There are four digits and the maximum reading, 1000 µl or 1 ml, is shown here.

Fig 5. This is a P1000 from a different manufacturer. This setting reads 941 µl.

This is a LOCK. Make sure it is in the UP position. If it is down, you will not be able to turn the dial without breaking gears of the micropipette.

Practice 1:

What is the minimum setting on a P-200 micropipette? _______

What is the minimum setting on a P-200 micropipette? _______

Practice 2:

Is the instrument shown here a P1000, P-200, or P-20? _______

As shown on the dial in the figure, what volume of liquid is this instrument set to deliver? _________

D. *Micropipette operating instructions*

1. Choose the micropipette appropriate for the volume you want to dispense. Looking at the volume indicator window turn the volume adjustment dial to the desired volume amount to be dispensed.
2. Attach the tip to the pipette immediately before use. Open the tip box and gently, but firmly, insert the pipette into the appropriate tip. <u>Close the tip box</u>!
3. Depress the plunger to the **first** stop before you go into the liquid that you want to dispense (do not do this while the tip is in the liquid).
4. Insert tip into the fluid to be dispensed; try to keep the pipette as vertical as possible and keep the <u>tip just under the surface</u> of the liquid.
5. Slowly and smoothly (you should not hear anything) release the plunger to draw up the fluid.
6. Remove the tip from the fluid; make sure that there are no pockets of air in the tip. With experience, you will be able to visually tell if the tip contains ≈ the right amount of fluid.
7. Place the tip into the receiving vessel; keep the pipette vertical and hold the receiving vessel at an angle so you can deposit the liquid on the vessel wall. Depress the plunger to the first stop.
8. To ensure that all the fluid is dispensed from the tip, depress to the **second** stop. Here it is especially important when working with very tiny amounts of fluid (*i.e.* 0.5-10 µl) to place the tip on the side of the vessel. This allows you to see that all the liquid gets deposited from the tip.
9. Remove the tip from the receiving vessel. To eject the tip, place the pipette above the tip waste container and press the ejector button.
10. To get all the liquid to the bottom of the vessel (Eppendorf tube), subject vessel to a brief centrifugal pulse in the microfuge.

Note: When accuracy is important, keep in mind that more viscous fluids will adhere to the sides (inside and outside) of the plastic tip, especially when working with tiny volumes (10 µl or less).

Practice 3A:

1. Select the appropriate instrument (P20, P200, or P1000) to pipette 450 µl.
2. Put the numbers – 4 – 5 - 0 – in the correct positions in the boxes of the appropriate instrument.

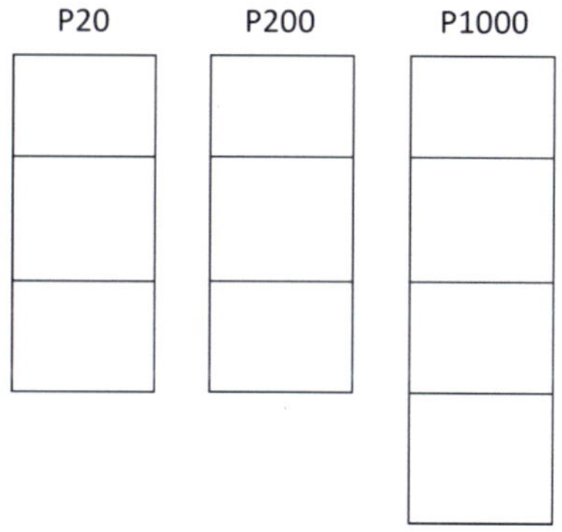

Practice 3B:

1. Select the appropriate instrument (P20, P200, or P1000) to pipette 11.8 µl.
2. Put the numbers – 1 – 1 - 8 – in the correct positions in the boxes of the appropriate instrument.

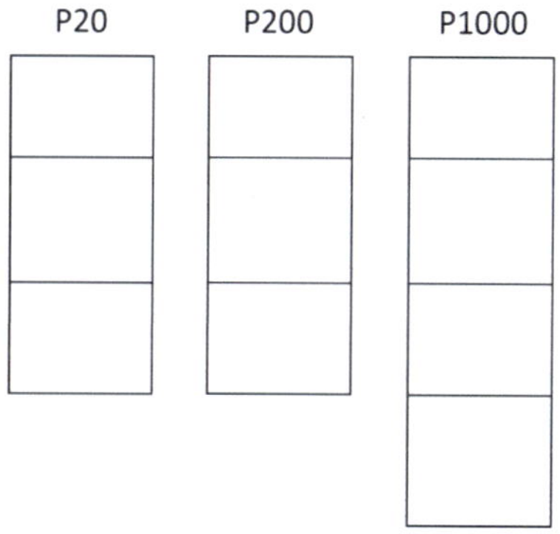

Practice 3C:

1. Select the appropriate instrument (P20, P200, or P1000) to pipette 187 µl.
2. Put the numbers – 1 – 8 – 7 – in the correct positions in the boxes of the appropriate instrument.

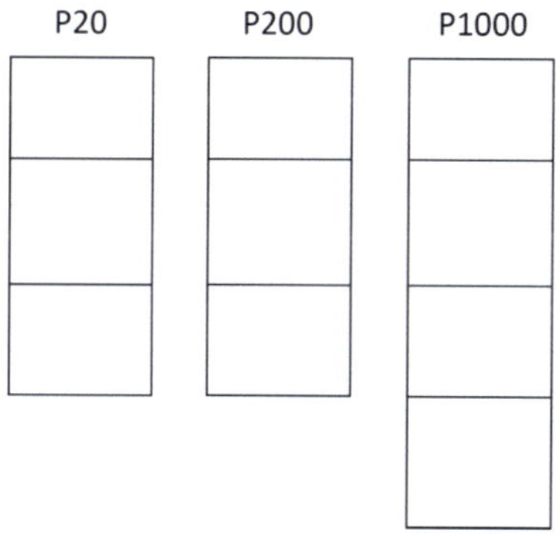

E. *Troubleshooting*

Problem	Possible reason	Possible solution
Air bubble in tip	Tip may have been just at the surface of the liquid – some air was drawn into the tip Plunger may have been released too quickly when filling the pipette	Tip should be submerged into liquid Release plunger more slowly
I pipetted the same volume to 10 tubes, but they do not appear to contain the same amount of liquid	Plunger was pressed to second stop while drawing liquid into the tip Air bubble in tip Tip not submerged in liquid	Repeat using proper technique
Tip does not stay on pipette shaft	Wrong size tip Tip not attached firmly enough	Use the correct tip Try a different pipettor Try a different box of tips; some tip brands work better for certain pipette brands
The dial won't turn on the micropipette	You have reached the limit of the instrument	Turning the volume dial too far can damage the instrument. You may need to repair or replace this micropipette.

9. ASEPTIC TECHNIQUE

A trained scientist must be able to determine which items/solutions require sterilization, understand the importance of timely sterilization of microbial growth media, identify the steps of aseptically inoculating a microbial culture, liquid and solid medium, and recognize microbial contamination.

Scientists and medical professionals must take care not to contaminate samples. Microbes are ubiquitous (everywhere). The term 'microbes' refers to live organisms including bacteria, archaea, fungi, protozoa, and algae, and non-living viruses . When working in a medical setting aseptic (sterile) technique is a strict requirement, as it is in most research labs. Aseptic technique is used to ensure that only the intended organism is cultured and that samples do not become contaminated with unwanted organisms.

Fig 1. The convection current produces a zone of safety around a flame that reduces the chances of microbial contamination.

THE DIFFERENCE BETWEEN DISINFECTION AND STERILIZATION

Disinfection is the **removal of many microorganisms from a surface** intended to greatly reduce the number of pathogens. Common disinfectants include 70% alcohol or 1% bleach.

Sterilization entails **complete elimination** of all living organisms. Growth medium (plural is media) and labware (bottles, tubes, tips) are sterilized by autoclaving in a pressurized chamber at 121°C, for >15 min (saturated steam is generated). Sometimes **Tindalization** (prolonged dry heat) or **chemical treatment** is sufficient to sterilize.

A. *Aseptic technique tips*

- Keep your workspace clean; before you begin, the workbench is disinfected using 70% ethanol or 1% bleach
- Create a sterile workspace on your bench by using a gas or alcohol burner. Use the flame to heat a metal loop or to heat the surface of an open tube during liquid transfer.
- Recognize items that are sterile. Sterile broth (liquid) medium or agar (solid) will be marked with autoclave tape, which develops a black stripe during the 15-20 min autoclave cycle (Fig. 2). Even if a solution has autoclave tape or has a *filter-sterilized* note, it is always possible that someone else was sloppy when they used the solution and introduced contamination. Try to control for this by obtaining small aliquots for your personal use.
- Sterile medium or solutions must be poured into a sterile container, such as a sterile Petri plate, sterile, capped test tubes, or beakers/flasks with foil.
- Don't introduce contamination into your sterile medium/solutions.
- Sterile beakers, flasks, or cylinders will be capped or covered with foil and labeled with autoclave tape (see Fig. 3) — uncapped beakers and graduated cylinders are **not** sterile.

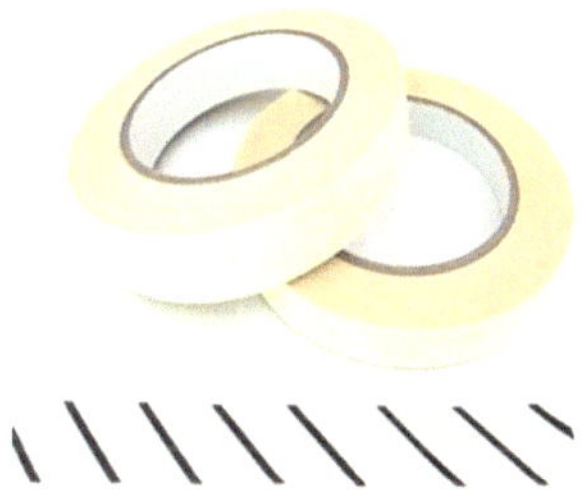

Fig 2. Autoclave tape turns black after an autoclave cycle

Fig 3. The dark autoclave tape on top indicates that this flask has been autoclaved.

Practice 1:
If a sealed flask has a darkened autoclave tape seal on it and no one has opened the flask, you can be positive that it is sterile.

True False

B. *Aseptic technique: pouring agar medium and working with Petri plates*

1. To prepare an agar plate, you must first autoclave medium (such as tryptic soy, LB, potato dextrose *etc.)* that has been supplemented with agar, typically 1.5% agar.
2. The autoclaved agar medium is cooled to about 50°C then poured into sterile Petri plates as shown in Fig 4. The 50°C bottle should be warm but not too hot; *don't insert a thermometer to check temperature as this will contaminate the medium.*
3. You will need to keep the Petri plates on a bench at room temperature until the agar has solidified (≈30 min).

Practice 2:
It is best to pour molten agar into Petri dishes immediately after removing the agar from the autoclave.

True False

Fig 4. Pouring sterile agar medium into a plate. Gently tip the bottle containing molten agar and pour in agar until the bottom on the plate is about half full of agar (≈ 25 ml).

C. *Culturing microbes on agar medium*

1. Once the agar has solidified, you can inoculate the plate. The streak plate method is a common way to inoculate the agar. This method allows you to obtain isolated microbial colonies that grow from a single, isolated cell.
2. The source material (source of microbes) for your agar plate is called the **inoculum**. It might be a broth culture or a colony from a plate from your instructor, a slurry from soil, your skin, etc.
3. A streak plate is generated by using flame-sterilized inoculating loop (metal or glass) or a sterile wooden stick (Fig. 5) to streak the inoculum onto a section of the agar plate.
4. Refer to the streak plate cartoon in Fig. 6. After the initial streak, the loop (rod) is flamed, cooled briefly, then touched to the surface where the initial inoculum was placed. The purpose here is to "pick up" some of the cells from the initial inoculum and streak them onto a new region of the plate.
5. Repeat step 4
6. After incubating the plate for 24-48 hrs (typically), colonies will be visible as shown in Fig. 6.

Fig 5. Sterile wooden inoculating sticks; the used sticks can be collected and reused after sterilization.

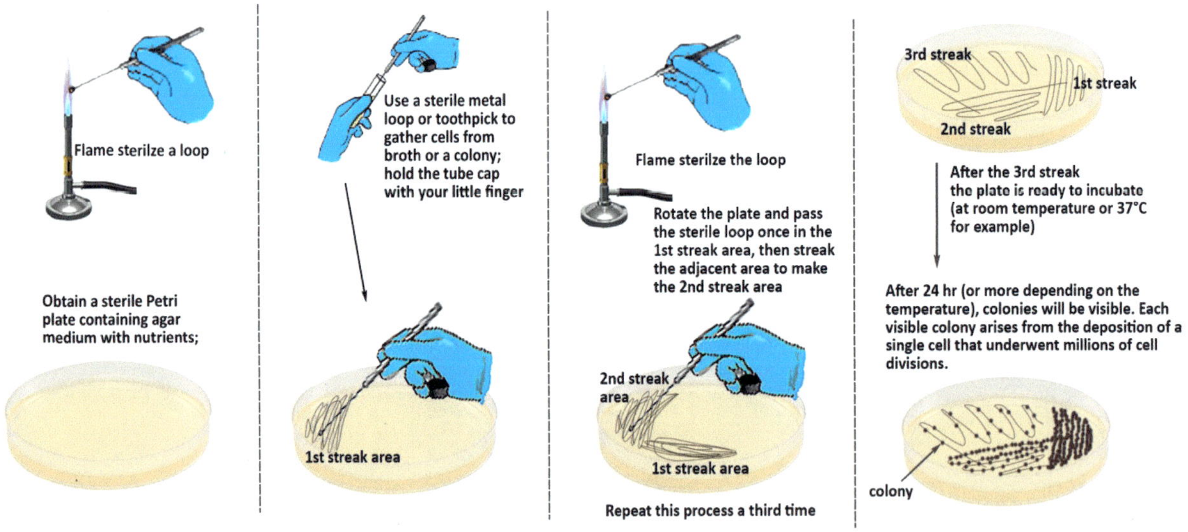

Fig 6. The drawing above illustrates the sequence of preparing a streak plate. When you use a sterile loop to streak a new section of the plate, you are diluting the population of cells. With careful technique, the third streak should result in the growth of isolated single cells that can grow into colonies.

D. *Culturing microbes in liquid*

Microbial cells (bacteria, yeast and other fungi, algae) are often grown in small batches of broth (liquid with nutrients) medium (≈ 3-5 ml). Aseptic technique is needed to transfer cells into sterile medium in containers (tubes or flasks).

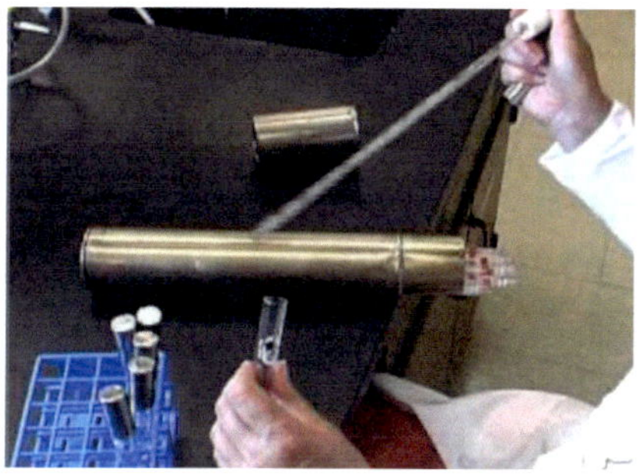

Fig 7. Use a sterile pipette (shown here) or a micropipette with a sterile tip to transfer sterile broth into a tube or flask.

How to inoculate liquid medium with cells suspended in broth

Transfer microbial cells from an existing broth culture using one of these instruments:

- a sterile glass pipette (shown in Fig 7)
- a heat-sterilized metal loop or glass rod (Fig 8)
- a sterile wood stick or toothpick [Note: do not flame the wood stick]
- a micropipette with a sterile tip

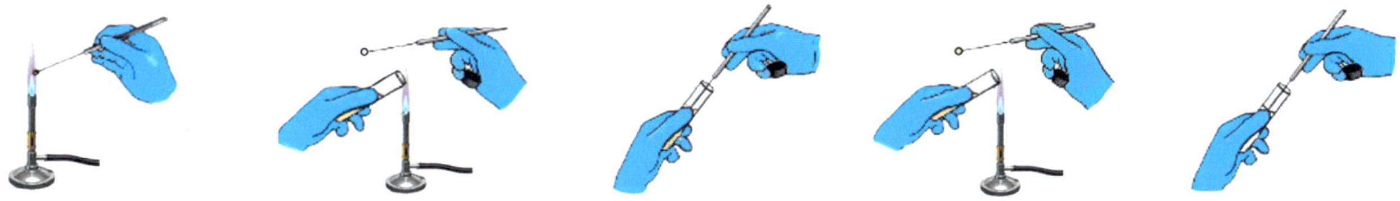

Sterilize the instrument (heat the metal loop or dip glass rod in 80% ethanol and insert into the flame).	Use your small finger (pinky) to remove and hold the cap of the tube containing the cell culture. **	Carefully insert the cooled loop (*a HOT loop can kill the cells*) into the liquid and swirl gently.	Pass the top of the tube containing the cell culture briefly through the flame and re-cap the tube.	Insert the metal loop (containing the cells – the ***inoculum***) into a new tube containing sterile broth.

Fig 8. Using a sterile inoculating loop to transfer microbes from one medium to another.
*** If you are taking cells from a plate, lift the lid of the Petri dish and touch the sterile loop to the colony.*

E. *Spread plating technique*

In addition to the streak plate method mentioned earlier, you can obtain isolated cells using the spread plate method. Whereas the inoculum for the streak plate can be from cells on a surface (a colony on an agar plate, your skin, your phone, etc.) or cells from a broth culture, the inoculum for the spread plate is almost always from cells in liquid (*i.e.*, the *inoculum*).

As shown in Fig 9, the inoculum is spread onto the agar surface with a sterilized glass "hockey stick" that is moved in circular movements. Alternatively, addition of sterile medium (at 50°C) directly on top of the inoculum will also distribute the cells over the agar surface.

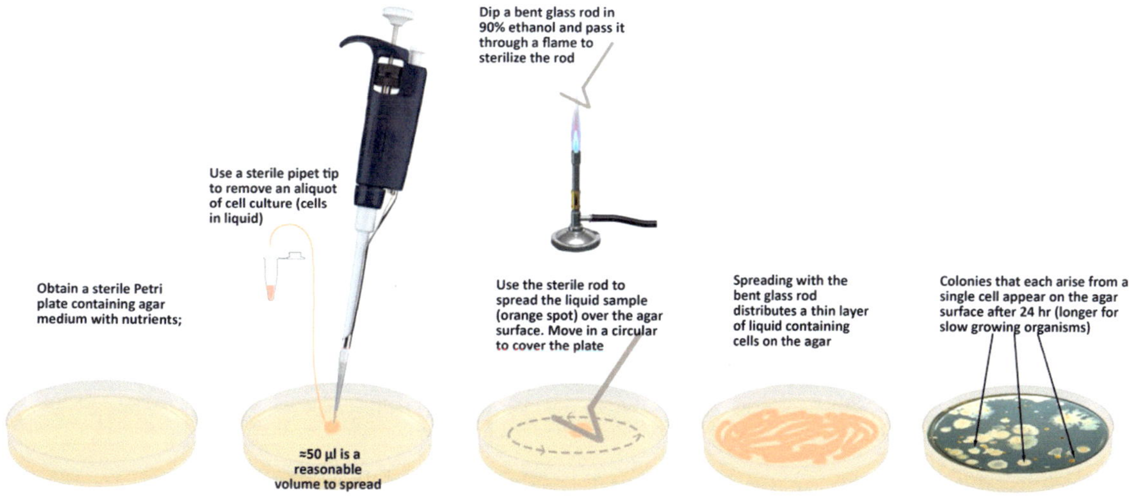

Fig 9. The spread plate method is used to place cells that will grow into visible colonies on agar medium. An aliquot of microbes can also be added to 50° agar medium to create a "pour plate" (not shown). Note that color has been used in this drawing for the purpose of visualizing the samples.

If the volume of liquid that is spread contains no more than 200 cells, each well isolated cell will grow into an isolated visible colony during incubation. If the liquid inoculum contains more than 500 cells the growth after incubation will not result in individual colonies, but a lawn of cells covering the surface of the plate.

COMPARISON OF STREAK PLATE VS SPREAD PLATE METHODS

Fig 10, left: a **streak** plate – cells are distributed on the solid medium using an inoculating loop or stick.

> Purposes: used to maintain cultures (weekly feeding), to separate two or more different organisms, and to check for contamination (assurance of a pure culture).

Fig 10, right: a **spread** plate – cells are distributed by spreading diluted cultures onto the solid surface of the agar.

> Purposes: used to separate two or more different organisms, and to enumerate cells.

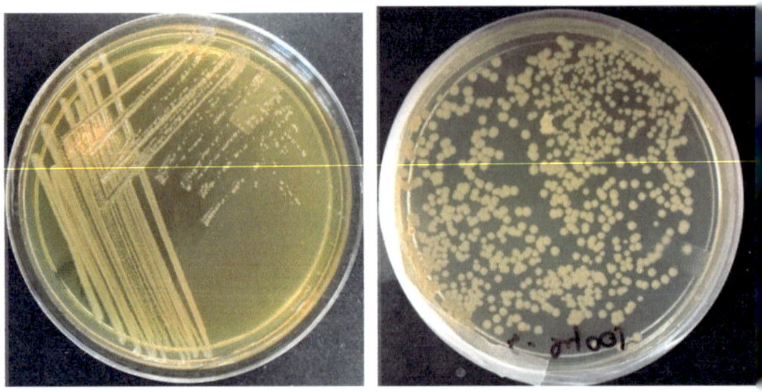

Fig 10. Left = a streak plate, inoculated with a sterile loop; Right = a spread plate, inoculated with a small volume (50-100 µl of a dilute bacterial culture.

F. *Troubleshooting*

Problem	Possible reason	Possible solution
Fuzzy growth on plates	Airborne contamination (perhaps due to fungal or bacteria spores) Note: when isolating microbes from soil, fuzzy growth may be normal on medium such as R2A, which can support *Streptomyces*	Take care when working with Petri plates; don't leave the lid off – lift it only briefly
Spreading growth on R2A plates	Some microbes, such as *Bacillus mucoides*, can swarm over the agar. 	Dilute your samples (e.g. soil) 1:100 or more and re-plate on new R2A medium
Unexpected growth on plates	Contamination introduced by poor aseptic technique	Be mindful of touching sterile items with hands/clothing Clean bench with bleach or 70% ethanol before starting work Work close to a flame Keep your tips sterile by closing the lid of the tip box
Liquid medium looks turbid	Medium may be contaminated or a precipitate may have formed due to added chemicals, such as a metal; place in the medium in the "to be autoclaved" waste tray	Obtain a fresh bottle of medium and use proper aseptic technique If you suspect a precipitate due to an additive, double check your calculations before making new medium
Nothing grew on my agar plates	You used the wrong medium You inoculating loop was too hot You did not inoculate The incubation temperature was too high The incubation time was too long/short	Make sure that you use the correct medium; *e.g.*, yeast like PDA or YPD, *E. coli* grows well on LB or NA

10. DILUTIONS AND WORKING WITH CELL CULTURES

A working knowledge of dilutions is essential for a scientist, particularly if you are planning a career in the medical field. An incorrect dilution can cost a life. Dilutions are also crucial when working with microbial cultures, in research or a environmental sample. Liquid (broth) cultures of microbes, such as yeast, algae and bacteria, can grow to very high cell densities. For example, a culture of the bacterium *Escherichia coli* can grow to a density of 1 x 10^9 cells/ml in 18 hours at 37°C (≈98.6°F). It is impossible to count cells at these high densities, so scientists use dilutions to achieve concentrations of cells that can be managed.

Terms
- **Solute:** the item being dissolved, such as a chemical
- **Diluent:** the liquid that the cells or chemicals are diluted into. Another term to describe diluent if solvent (used more in chemistry)

Learn to

- Aseptically carry out simple dilutions and serial dilutions
- Spot aliquots of each dilution onto an agar medium and monitor growth of colonies, such as yeast or bacteria
- Quantify the concentration of cells in a broth culture after seeing that single, viable cells can grow into a visible colony in a few days. Depending on the type of organism, colonies should appear in 2-4 days at room temperature (22°C). Many organisms exhibit faster growth at higher temperatures, such as 30° or 37°C.

A. *Simple dilutions for chemical solutions or microbial cells in culture medium*

Simple dilutions combine defined amounts of the sample being diluted (volume of cells or solute) and a diluent (or **solvent**) as shown in the table below.

$$\textbf{DILUTION FACTOR} = \frac{\textit{the volume of the sample being diluted (cells or chemical solute)}}{\textit{TOTAL volume (the volume of the sample being diluted PLUS the volume of diluent)}}$$

Volume of sample (cell cultures or chemical solutions)	1000 µl	800 µl	600 µl	400 µl	200 µl	100 µl	50 µl
Volume of diluent (or solvent)	0 µl	200 µl	400 µl	600 µl	800 µl	900 µl	950 µl
TOTAL VOL	1000 µl	1000 µl	1000 µl	1000 µl	1000 µl	1000 µl	1000 µl
Dilution factor (see equation above)	$\frac{1000}{1000}$	$\frac{800}{1000}$	$\frac{600}{1000}$	$\frac{400}{1000}$	$\frac{200}{1000}$	$\frac{100}{1000}$	$\frac{50}{1000}$
Simplified dilution factor	1:1	4:5	3:5	2:5	1:5	1:10	1:20

Here are two examples of an experiment where you would use simple dilutions:

1. To dilute a culture of bacterial cells that will be placed in the spectrophotometer to read absorbance.
2. To dilute a chlorophyll sample that you extracted from plant tissue to determine the concentration of chlorophyll (using the extinction coefficient for chlorophyll).

Practice 1:
Fig 1 (right) shows dilutions of a blue compound. Circle the tube that contains the most concentrated blue solution.

Practice 2:
You have 500 µl of cell culture added to 4.50 ml of diluent (solvent).

What is the total volume of the solution? ____________

What is the dilution factor (the simplified dilution factor)? ______________

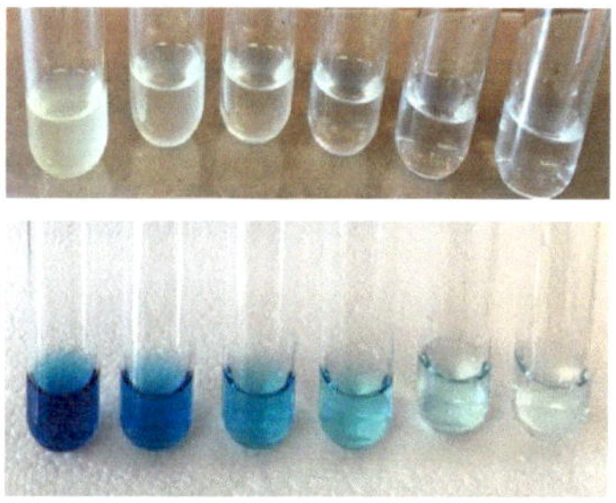

Fig 1. The top panel shows six dilutions of bacterial cells, with the most concentrated sample on the left. Dilutions of a blue compound are shown in the bottom panel.

B. *Serial dilutions of concentrated solutions, including microbial cell cultures*

Microorganisms multiply readily to produce high cell densities. For example, *Saccharomyces cerevisiae* and *Escherichia coli* (in the proper growth media) can grow overnight to produce as many as 10^7 to 10^{10} cells per milliliter, respectively. Best practice is to dilute these concentrated cell cultures *via* **serial dilution** (Fig 2).

Serial dilutions are preferred over simple dilutions when you need to make dramatic changes in concentration. A serial dilution refers to sequential dilution steps —this avoids creating large volumes of the desired dilution.

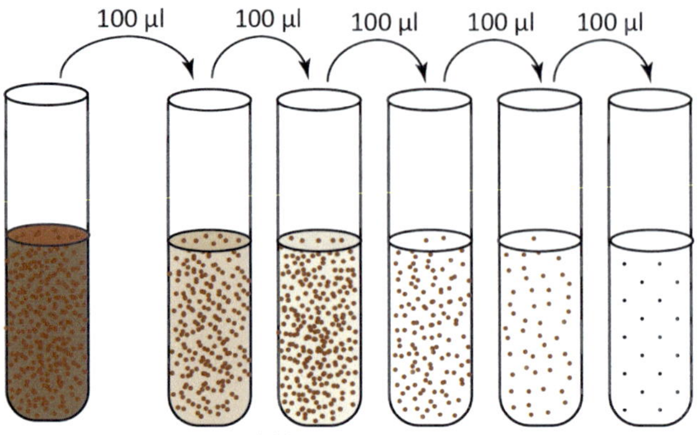

Fig 2. Serial dilutions are used to reduce the concentration of cells in a solution until a manageable concentration is achieved.

EXAMPLE

What volume of water would you need to dilute 1 ml of a culture containing 1 x 10^8 cells/ml to 1 x 10^2 cells/ml?

Start with 1 x 10^8 cells/ml = 100,000,000 cells/ml

Volume = 1 ml

Desired final concentration = 1 x 10^2 cells/ml = 100 cells/ml

Final volume = unknown

Try using $C_1V_1 = C_2V_2$

100,000,000 cells/ml x 1 ml = 100 cells/ml x V_2

On the left side of the equation, the ml units cancel

V_2 = 100,000,000 cell ÷ 100 cells/ml = **100,000 ml**

100,000 ml = 100 liters of liquid! (this is about 25 gallons)

Alternative method: use serial dilutions to dilute 1 ml of a culture containing 1 x 10^8 cells/ml to 1 x 10^2 cells/ml. Look at the serial dilution model in Fig 2. If the original sample contained 100,000,000 cells in 1 ml and you now add the 1 ml to 9 ml of buffer (this is a 1:10 dilution, $\frac{1}{9+1}$, also written as 10^{-1}) you have diluted the cell concentration as follows: 100,000,000 cells are now in 10 ml total volume, hence the concentration is 10,000,000 cells/ml.

$C_1V_1 = C_2V_2$

C_1= 100,000,000 cells/ml

V_1 = 1 ml

C_2 = ??

V_2 = 10 ml (total volume)

C_2 = 10,000,000 cells/ml → this is a 10-fold decrease in the cell concentration.

Each subsequent 10-fold serial dilution will further decrease the cell concentration.

C_1 = concentration of item 1
V_1 = volume of item 1
C_2 = concentration of item 2
V_2 = volume of item 2

To enumerate the precise number of cells, concentrated cultures are aseptically diluted several times until the concentration of cells is sufficiently reduced (*i.e.,* a countable number of cells) to allow enumeration. To enumerate, one must either count cells in a cell counting chamber using a microscope or plate diluted cultures on solid media. During incubation, if you successfully deposited single, viable cells (well separated), each single cell should grow into a visible colony. The cell number of the original culture can be determined by multiplication with the dilution factor.

Practice 3:
You have 5x10^7 cells/ml in your cell culture. What is the concentration (in cells/ml) of a 1:100 dilution? ____________________

Using serial dilutions to determine the number of cells in a culture

You are carrying out an experiment with *E. coli.* Your professor gave you a tube of cells, but you need to know the concentration — the number of cells/ml (also referred to as cell density)— before you can work. The tube looks quite turbid, so you suspect that the cell density is high – more than a million cells per milliliter (1 x 10^6). Densities of bacterial cells can reach 1x 10^{10} cells/ml overnight. A sample serial dilution is shown in Fig 1 (previous page) using 100 µl + 900 µl to achieve serial (consecutive) 1:10 dilutions. The example below illustrates the use of smaller volume serial dilutions followed by a spot plate to determine the number of cells/ml in your sample.

EXAMPLE

1. Perform serial dilutions: Fig 3 shows a 1:10 dilution series using smaller volumes, 100 µl + 900 µl, than shown in Fig 1.
2. For each dilution, you must use a fresh, sterile tip to transfer the sample.
3. Spot plate: as shown in Fig 3, after the dilutions are complete, an aliquot (typically use 2-5 µl) is transferred from each tube to a Petri plate containing sterile agar medium.
4. Incubate at room temperature (RT) or 30°C.
5. Check plates every day and note when growth is visible; see Fig 4.

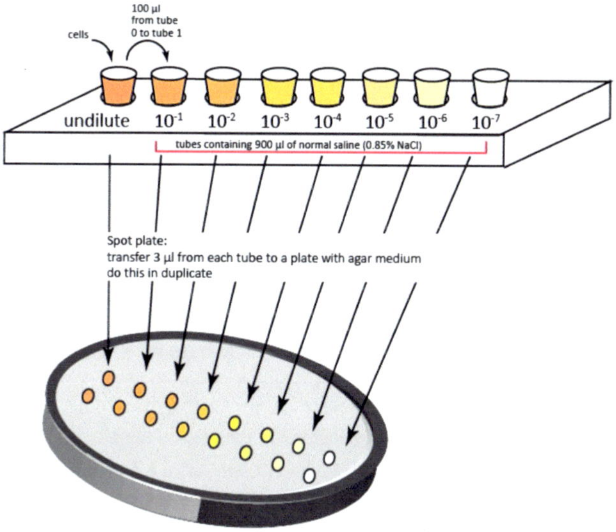

Fig 3. 10-fold serial dilutions are used to reduce the concentration of cells in a sample to a countable number. Aliquots of each dilution are spotted on agar medium and incubated to allow individual cells to form visible colonies. The colony that you see is an indication of the number of viable cells that were present in the aliquot and can be used to calculate cells/ml (cell density).

Practice 4:
Look at Fig 4. Each isolated white spot that you see is a colony that formed from a single yeast cell after incubation at 30°C for 48 hr. If the initial 3 µl aliquot contained many yeast cells, you will not be able to distinguish individual spots because growth from many yeast colonies overlap.
Count the number of colonies that you see and record your results in the table below. *If you cannot distinguish individual colonies, estimate the expected number of colonies based on the dilutions.* The EXAMPLE shows calculations that are based on a spot plate (not shown) and estimated from the dilutions.

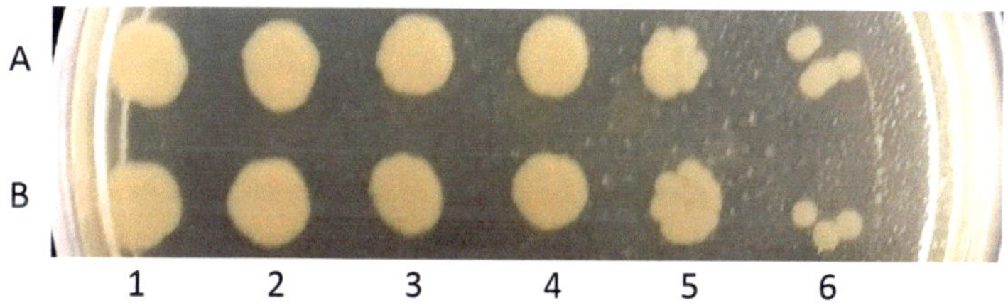

Fig 4. In duplicate (rows A and B), 3 µl aliquots of cells were spotted on this plate containing potato dextrose agar medium and the plate was incubated for 2 days at 30°C. From left to right dilutions are 1=10^{-1}, 2=10^{-2}, 3=10^{-3}, 4=10^{-4}, 5=10^{-5} and 6=10^{-6}.

row/dilution	Number of yeast colonies in this spot					
	1 (10^{-1})	2(10^{-2})	3(10^{-3})	4 (10^{-4})	5 (10^{-5})	6 (10^{-6})
HYPOTHETICAL EXAMPLE	600,000	60,000	6,000	600	60	6 (actual count)
The estimate of yeast colonies for each dilution above comes from the colony count in spot 6 and each serial 10-fold dilution. Look at Fig 4 and estimate the number of colonies for each dilution of sample A and B.						
row A (Fig 4)						
row B (Fig 4)						

Additional data analysis

1. The growth from the aliquots of cells that you spotted from each dilution is shown in Fig 4.
2. Find a spot that shows isolated colonies (Fig 4, #6).
3. Count the number of colonies in the spots where you can see distinct colonies — each colony arose from an individual, isolated cell.
4. Examine the colony growth in Fig 4. Each white spot is a colony (MANY cells) that grew from a SINGLE yeast cell. The more concentrated samples have so many cells that you see one continuous white circle.
5. You count 5 and 3 colonies in the two spots that came from the 10^{-6} dilution.
6. Calculate the average: 5+3 = 8 → 8÷2 = average of 4 colonies per spot
7. You originally spotted 3 µl of each dilution.
8. Each isolated colony probably arose from a single cell, so if there are 4 colonies, there must have been 4 viable cells in the 3 µl aliquot that you plated initially.
   Hence, 4 colonies/3 µl = 1.3 cell/µl in the 10^{-6} dilution.
9. Next, extrapolate back to the original sample
   - If the 10^{-6} dilution contained 1.3 cell/µl, then
   - the 10^{-5} dilution had 10 times more cells per µl or 13 cells/µl
   - the 10^{-4} dilution had 10x more cells per µl or 130 cells/µl
   - the 10^{-3} dilution had 10x more cells per µl or 1300 cells/µl
   - the 10^{-2} dilution had 10x more cells per µl or 13000 cells/µl
   - the 10^{-1} dilution had 10x more cells per µl or 130,000 cells/µl
   - the original (undiluted) sample (not included in the spot plate) contained 1,300,000 cells/µl
   - report this as cells/ml so multiply by 1000 to covert µl to ml

The original culture contained 1,300,000,000 cells/ml or 1.3 x 10^9 cells/ml

C. *Troubleshooting*

Problem	Possible reason	Possible solution
Nothing grew on the 10^{-3} dilution, but there were colonies on the 10^{-4} dilution on your spot plate	You may have forgotten to spot from the 10^{-3} dilution tube; alternatively, perhaps you skipped the 10^{-3} tube when you were preparing the dilutions	Redo the dilutions and spot plate if you have time Tip: to keep track of which tube you have taken a sample from, move the tube after you are done using it
All spots resulted in a "lawn" of growth	You forgot to dilute the cell culture You did not dilute enough Your dilution calculations were wrong Your made errors while pipetting	Original culture was more concentrated, *i.e.* more cells/ml than anticipated → repeat and add additional serial dilution steps
Spots grew into each other and are not round; spots are runny	Plates were too fresh (or wet after refrigeration) or aliquots spotted were too large	Have plates at room temperature before use; make sure the surface is dry. Repeat using drier plates, do not spot more than 5 µl aliquots
No growth on spot plates	You set up dilutions but did not actually use the initial culture that contained cells; *i.e.*, your spots contained only sterile water	Repeat the dilutions; make sure you use the culture to make the first dilution.

11. MICROSCOPY

Microscopes are able to <u>resolve</u> items (reveal detail) that we cannot see with the human eye.
General science lab microscopes come in two basic designs:

- **THE STEREOMICROSCOPE (*ALSO* CALLED A DISSECTING MICROSCOPE)**
  (magnification range ≈ 2X to 50X)
  Used for looking at plants, insects, tissue, details of small organisms, complex samples (soil)
- **BINOCULAR MICROSCOPE**
  (magnification range ≈ (40X to 1000X)
  Used for examining microbes (protozoa, fungi, bacteria *etc.*) and cells.

A. *The stereomicroscope*

Samples for the stereomicroscope ((shown here is the Leica EZ4) are typically placed on a glass plate or a in Petri plate on top of the LED transmitted illumination source for viewing.

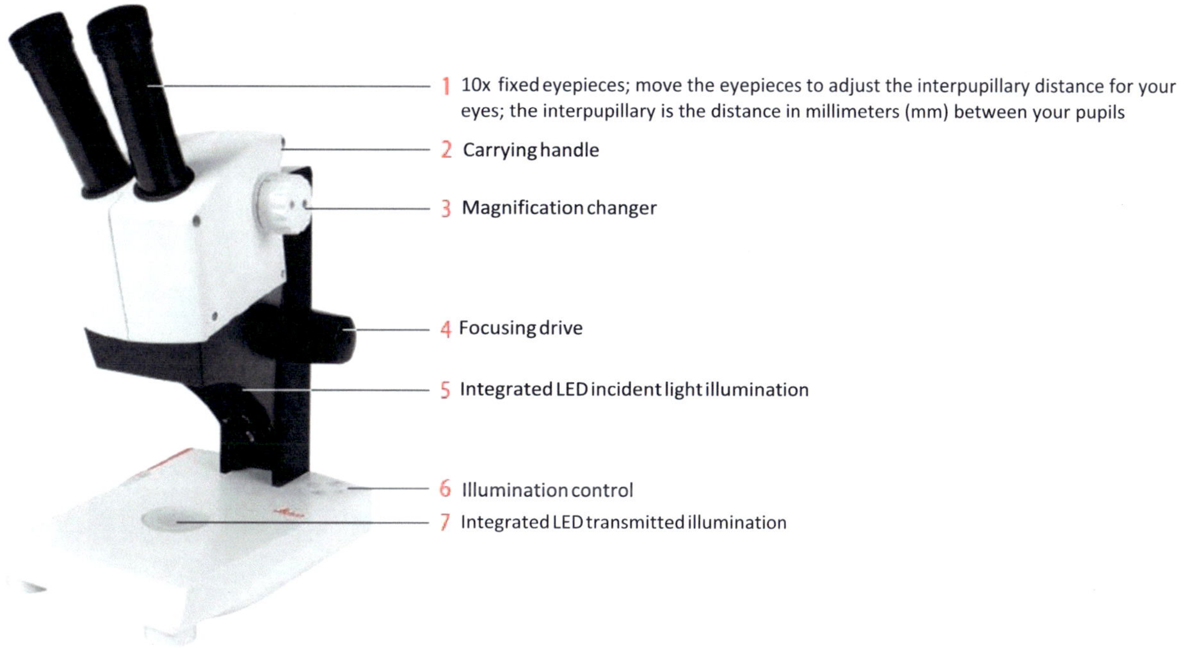

Fig. 1 A typical stereomicroscope (also called a dissecting microscope)

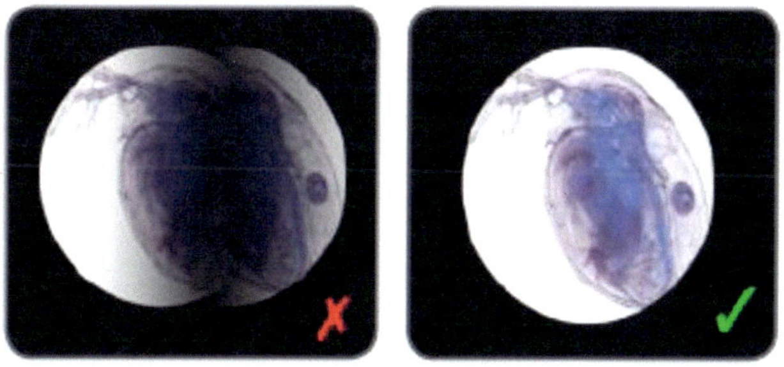

Fig 2. Adjust the interpupillary distance until you see a single image.

B. *The binocular microscope*

Example: the Leica DM500 phase microscope

Samples for the binocular (bi = two, ocular = eyepieces) microscope are placed on a glass slide for viewing. If the sample is wet, a cover slip must be placed on top of the sample to protect the microscope lens from damage and contamination.

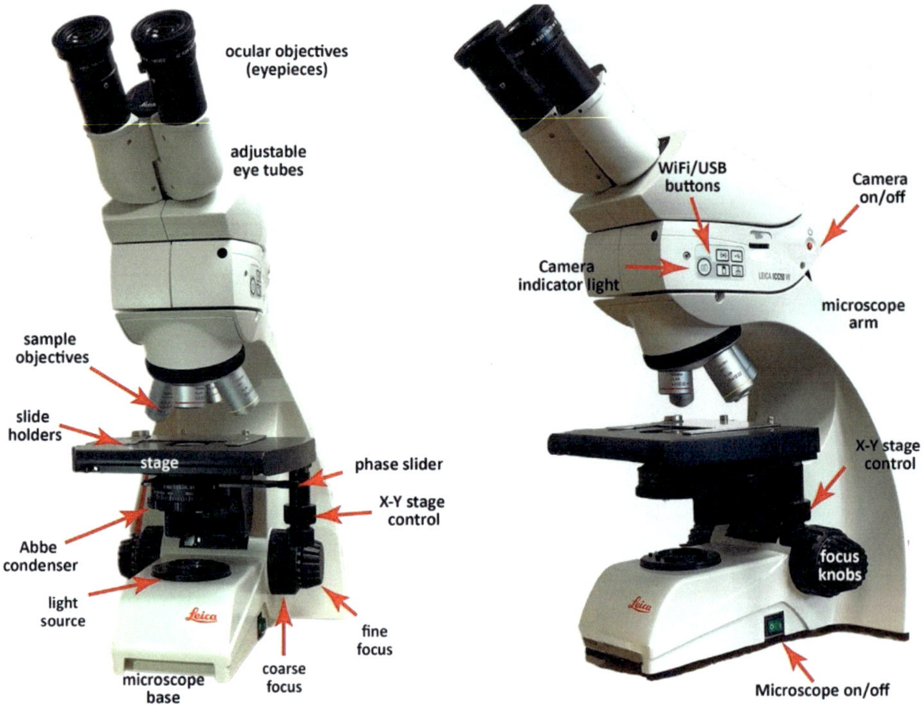

LEARN THESE PARTS OF THE BINOCULAR MICROSCOPE

- on/off switches for microscope and camera
- fine and coarse **focus knobs**
- light source (and field iris diaphragm control)
- stage platform for your slide; use the X-Y stage control to move the stage
- **ocular objectives** (*aka* ocular lens and eye pieces)
- adjustable eye tubes for setting interpupillary distance
- **sample objective lens** –the lens closest to the sample
- the **revolving nosepiece** holds the sample objectives; sample objectives are ***parfocal***, which means that the focal plane is maintained when switching between different objective lenses. i.e., if you focus on an object with the 20X objective, the sample should still be in focus when you rotate to the 40x objective.
- **Abbe condenser lens**–it concentrates and controls the light that passes through the specimen

Fig. 3. Move the eyepieces so that they align with your eyes

Samples need to be on a glass microscope slide.

A glass cover slip needs to be on top of the sample – this will prevent the microscope lens from being contaminated.

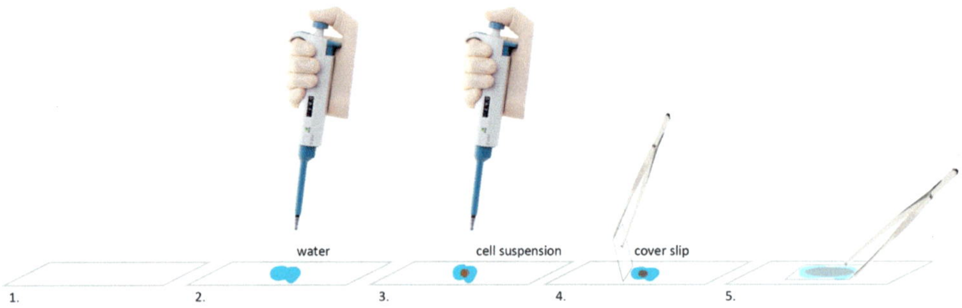

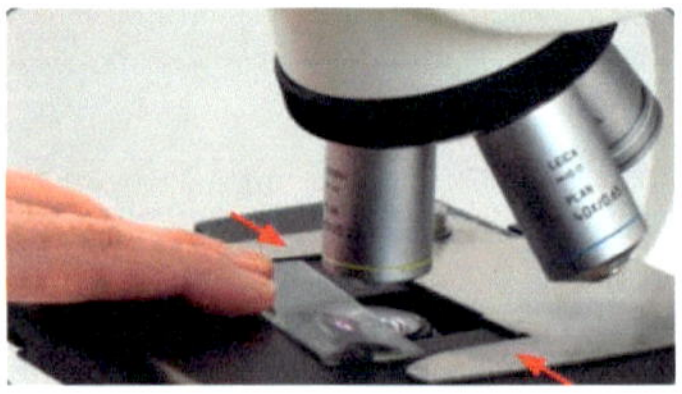

Slide holders

Fig. 4. Preparing a wet mount. Left to right: 1. Put a clean glass slide in a surface. 2. Add ≈40 µl of water or buffer. 3. Add ≈10 µl of cells (from broth culture or cells from a colony). 4. Use forceps to place the edge of a coverslip in contact with the edge of the water drop. 5. Lower the coverslip onto the water/buffer + cell mixture to form a thin film of liquid. Photo: Place the specimen slide in the microscope stage between the slide holders (red arrows).

C. *Correct operation of the microscope*

1. Plug in the microscope to a power source
2. Turn the microscope on — the **on/off switch** is at the bottom right of the microscope [it may be in the back of some microscopes]
3. Check the illumination intensity dial and set to the lowest value.
4. Move the smallest sample objective (often a 4x or 10x lens) to the front position (closest to you) of the revolving nosepiece
5. Place specimen in the slide holder on the stage; the slide grips will hold the slide firmly in place (see Fig. 3 above)
6. Use the X/Y stage control knob to position the specimen under the objective (still using the lowest magnification)
7. Adjust the interpupillary distance of the ocular (eyepiece) objectives to fit your eyes
8. Set the condenser and the phase slider to MATCH the objective in use
    a. Adjust the Abbe condenser to **<u>match the setting</u>** indicated on the sample objective (see closeup images in Fig. 5)
    b. Gently pull the phase slider on the left side (beneath the stage) to match the sample objective

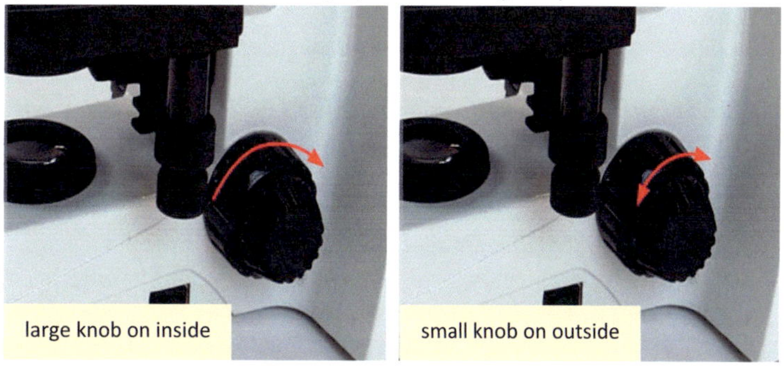

Fig. 4. There are two knobs too adjust focus; the large knob is the coarse adjustment (use this first); the small knob is the fine adjustment

9. Adjust the lighting to a level comfortable for your viewing
10. Slowly begin to turn the coarse focus knob (the large knob closest to the microscope body) until you begin to see your sample
11. Switch to the fine focus knob (the smaller knob connected to the coarse focus knob); turn slowly until your sample is in focus — the sample should look sharp and the contrast should be high
12. As you move to higher objectives be sure to change the phase slider and Abbe condenser settings to <u>match the objective</u>
13. **After you are done …**
    - Clean the microscope; use <u>lens paper</u> on the microscope objectives (*all other paper will scratch the lenses*).
    - Always carry the microscope using the handle while supporting the base

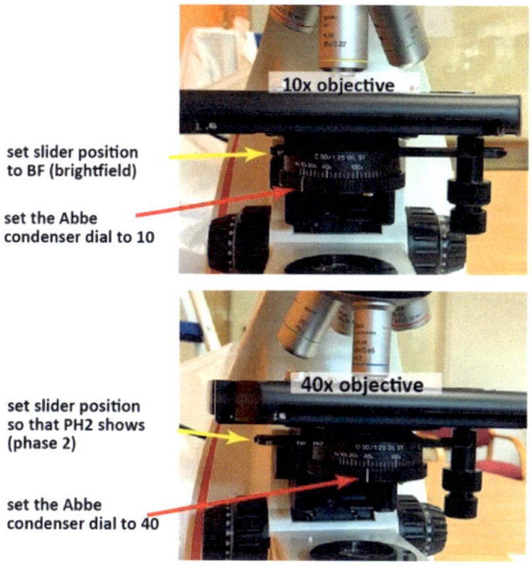

Fig. 5. Examples of the proper configuration for the 10x and 40x objectives.

TOTAL MAGNIFICATION AND SIZE BARS

It is critical to report the <u>total magnification</u> when you are communicating your results to others. In most science publications, **size bars** are required to show a line of known length (for example, a bar showing 10 μm). The total magnification and/or the size bar allow another person to determine the size of the specimen.

The **total magnification** is calculated by multiplying the magnification of the ocular objective (also called the eyepiece; it is typically a 10x objective) by the magnification of the sample objective lens. The magnification of the sample objective lenses typically ranges from 4x to 100x. for a total magnification of 40x to 1,000x, respectively.

Total magnification = ocular objective magnification × sample objective magnification

EXAMPLE: **10x × 40x = 400x**

D. *Troubleshooting*

Problem	Possible solutions
Image is fuzzy	Check your focus; make sure the Abbe condenser and the phase slider are in the right positions
No image	Make sure that the part of the microscope containing your sample is in line with the sample objective lens
Objective lens touches the slide	The lenses are parfocal. If you have focused on the incorrect image (such as an air bubble) or did not focus on an object using the 4X or 10X sample objective lens, moving to a higher power objective lens will not yield good results and may damage the lens
Poor contrast	Turn the light down

E. *Taking microphotographs*

1. Make sure the black cord at the back of the microscope is plugged into the top and the bottom – this sends power to the microscope. The microscope connects to an iPad *via* WiFi.
2. Turn the WiFi Fig 6, #2
3. Turn on the camera Fig 7, #5; the light above the camera button (#2) will turn green when the camera is ready
4. Go to Settings on the iPad (images in Fig 8) and open WiFi; connect to the closest microscope – the ICC code is located on the front of the microscope (either on the stage or just below the eye pieces)
5. Enter the password – leicamicro
6. On the iPad open the app Leica AirLab
7. Click on the camera image and select your microscope (not demo mode)
8. The iPad should display the image that you are looking at through your microscope
9. To take an image hit the red button on the top, right side of the tool bar.
10. You can send your photo to your email account, return to the SETTINGS app on the iPad and connect to your network. You can select your image and email it to yourself.

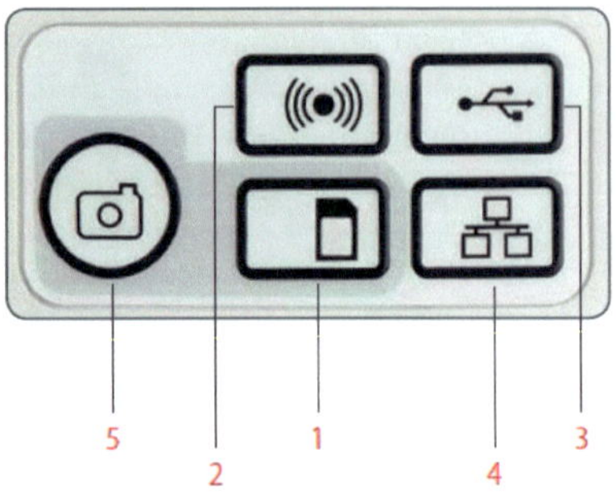

Fig 6. Camera connectivity options on the Leica microscope: WiFi (#2) to your phone or use a USB connection (# 3).

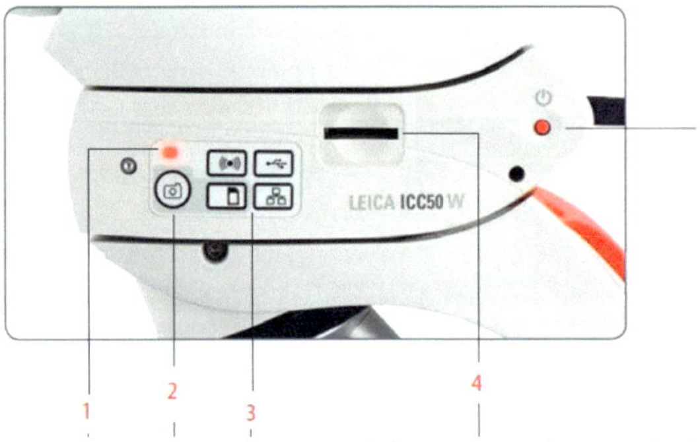

Fig 7. Leica camera indicators: #5 is the camera ON/OFF switch; #1 will turn green when the camera is ready.

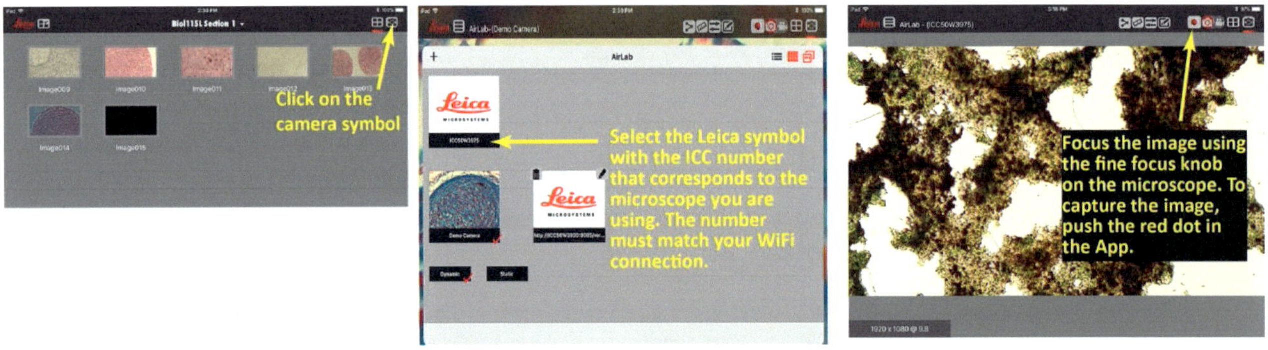

Fig 8. Sequence of screenshots for the camera app (from left to right). First, click on the camera symbol. Second, select the correct Leica microscope/camera. Third, take a photo of your focused sample.

Practice 1:
Draw an arrow to the Abbe condenser on the microscope image →.
What is the function of the Abbe condenser?

Practice 2:
You are looking at a slide of yeast cells using the 20x objective. What is the total magnification to your eyes?

Hint: 20x • the magnification of the ocular lens (look at the number on the eyepiece) (• = multiply) = ____________

F. *Resolving power*

The **resolving power** (Fig 9) of an objective lens is measured by its ability to differentiate two points that are close together. Resolution depends on the numerical aperture of the microscope objective (a number written on each objective) and on the on the wavelength of light used to illuminate the specimen.

The greater the resolving power, the smaller the minimum distance between two lines or points that can still be distinguished. The larger the numerical aperture (NA) of the objective, the better the resolving power (smaller resolution distance; Table 1). Resolution is given in units of microns (**µm** = micrometer or 10^{-6} meter).

Resolution (r) = λ/2 x NA

Table 1. Use the numerical aperture of the objective to determine the resolution.

Objective magnification	Numerical Aperture (NA)	Resolution
4x	0.10	2.75 µm
10x	0.22	1.30 µm
20x	0.40	0.69 µm
40x	0.65	0.42 µm

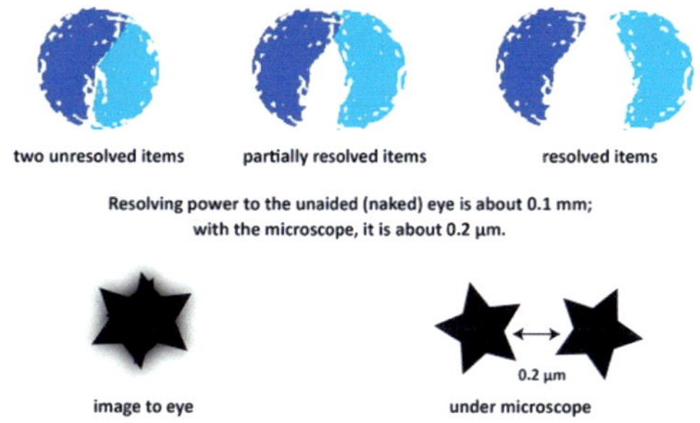

Fig 9. Resolution is the ability to distinguish two closely spaced objects as separate entities. A microscope can improve resolution about 500 times compared to the eye.

Examples of four different microscope objectives

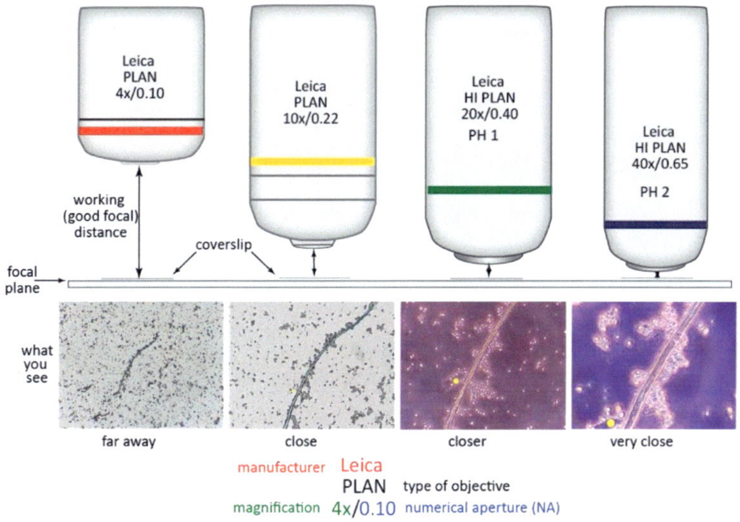

The four images above show the same sample – yeast cells and a single fungal hypha – at increasing magnifications; a yellow spot has been added to help you track the position. With increasing magnification, the ***resolution*** (*r*) of the image improves. The value ***r*** tells you the distance between two objects that the objective can distinguish.

The resolution can be calculated as follows:
r = wavelenght **(λ)/2x NA**

The NA is printed on the objective. Basic microscopes use white light (λ = 390nm - 700nm) to illuminate the speciman, so we can use ≈550nm (the average or mid-wavelength) in the equation.

For the 10x objective, ***r* = *550nm*/2 x 0.22 = 1250 nm or 1.3 µm**

For the 40x objective, ***r* = *550nm*/2 x 0.65 = 423 nm or 0.42 µm**

How to interpret these numbers: two cells viewed with the 10x objective will need to be at least 1.3 µm apart for you to be able to see two objects. With the 40x objective, they only need to be 0.42 µm apart for you to see each cell distinctly.

Fig 10. As the objective changes from 4x, 10x, 20x to 40x, the lens gets closer to the sample. At each step, the magnification and resolution increase.

Practice 3:

In the image here, what is the function of button #1?

What color is the #2 button when the microscope camera is ready to take photos?

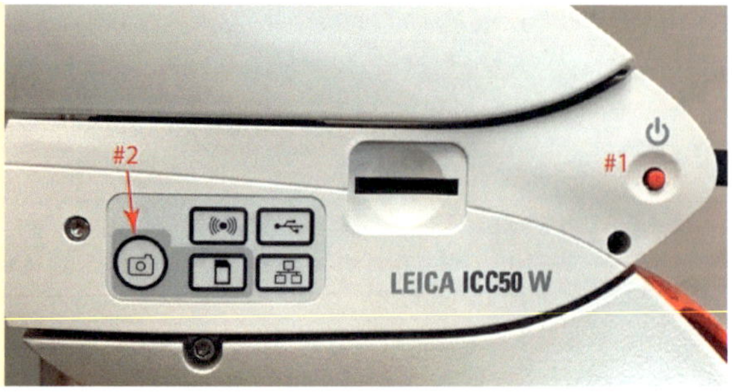

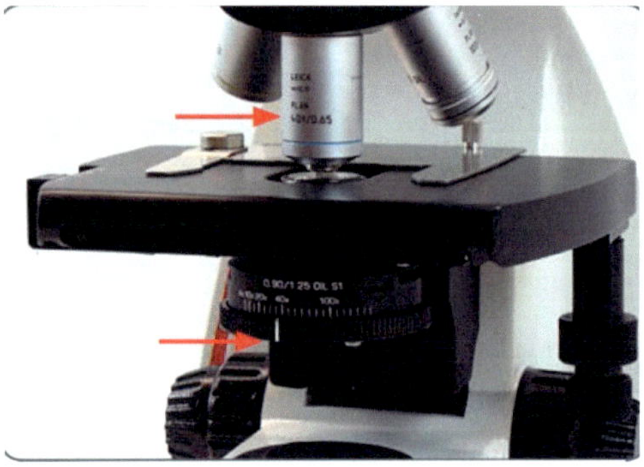

Practice 4:

If the 40x objective is in use, as shown here, what number setting should you use for the Abbe condenser?

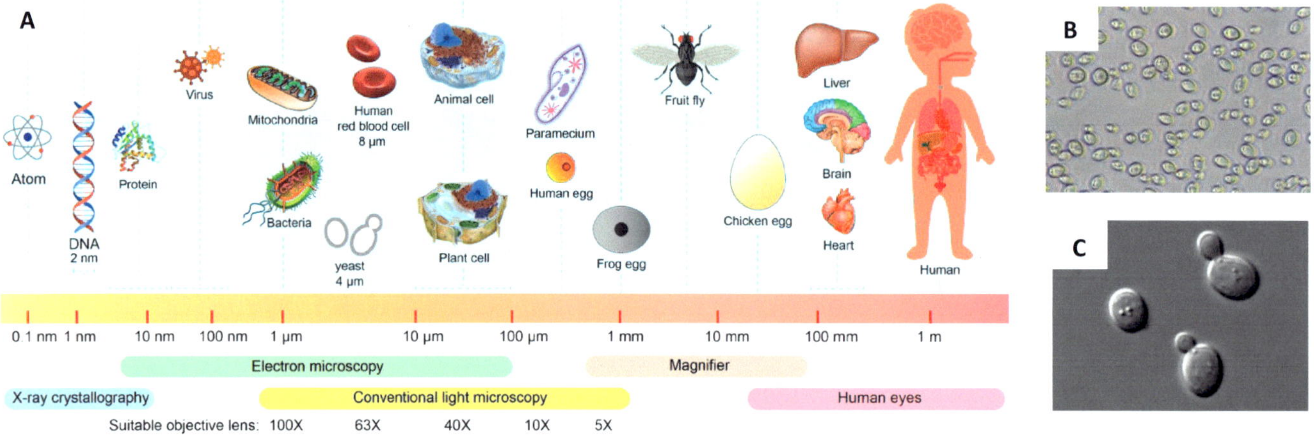

Fig 11. A. Cartoon showing a size comparison of some familiar items and microscopic images. B. Photograph of yeast cells at 400x magnification. C. Photograph of yeast at 1000x magnification.

12. USING SPECTROSCOPY TO ANALYZE CONCENTRATION

Scientists use spectrophotometers to identify chemicals in solutions, to calculate chemical concentrations (nucleic acids, protein, pigments) and to estimate cell concentrations. The spectrophotometer unit is called absorbance (A) because materials <u>absorb</u> particular wavelengths of light. The basic lab spectrophotometer uses light in the ultraviolet-visible range (UV-vis) whose wavelengths range from 230 nanometers (nm; 10^{-9} meter) to ≈ 800 nm. The term optical density (OD) is the unit used to describe cell solutions measured in a spectrophotometer (usually at a wavelength of 600 nm).

> Wavelength is sometimes abbreviated as λ (lambda). The wavelengths of light for spectroscopy include visible, ultra-violet, and infrared.

Visible light consists of different wavelengths of light the sum of which adds up to white light (Fig 1). We see the color of an object because the object reflects a particular wavelength of light (Fig 2). In the case of a solution, the observed color is determined by the wavelength(s) of light that is not absorbed, i.e., that can pass through the solution.

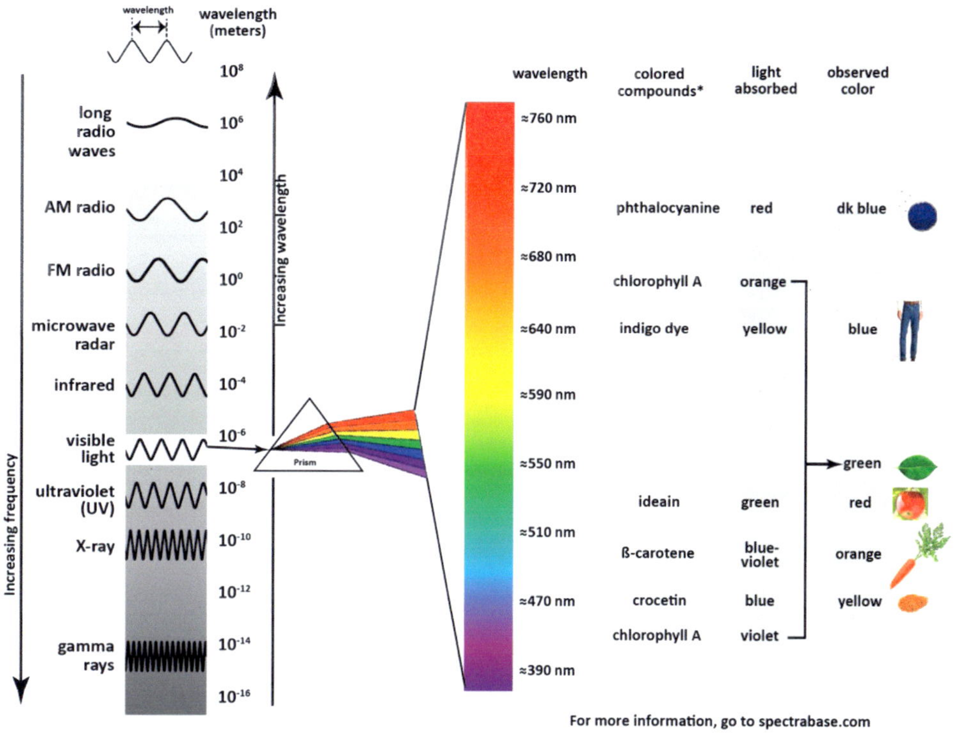

Fig 1. Humans take advantage of most of the electromagnetic spectrum, shown here. The left column shows the wavelengths associated with different spectral bands. Humans 'see' only a tiny portion of the spectrum –visible light – the colors for which were first revealed by passing light through a prism. The right columns show an expanded visible spectrum and examples of a few compounds that absorb particular wavelengths of light. A compound such as phthalocyanine absorbs red light and reflects other colors – it 'looks' blue to our eye.

A. *The design of a spectrophotometer*

A spectrophotometer consists of a light source, a prism and an instrument that detects (measures) the light as shown schematically in Fig 3. The slit, positioned after the prism, allows the operator to select a specific wavelength of light. If a sample in a quartz or plastic cuvette is placed between the light and the detector, the amount of light that reaches the detector will change because the sample will absorb some of the light. If the cuvette contains water, very little light will be absorbed. If the water contains a material (a compound or cells, for example) that absorbs more light than the water, the absorbance will increase. The higher the concentration of the light-absorbing solute, the more of the initial light is absorbed.

The information you get from a spectrophotometer, *i.e.*, the <u>output</u>, is a reading called <u>**absorbance**</u> (abbreviations: A or Abs). Note in Fig 3 below that absorbance is plotted on the Y axis. The linear range of *newer* instruments can range from 0-3 <u>a</u>bsorbance <u>u</u>nits (abbreviated as AU); 0-2 AU in older instruments. The linearity depends on the type of sample you are reading. As mentioned above, the spectrophotometer can be used to determine the concentration of a compound in a solution (absorbance) or the number of cells in a broth culture (optical density).

Fig 2. Sunlight is called "white" light, but it actually contains all wavelengths of the visible spectrum.
This apple looks red because it produces a pigment called ideain (an anthocyanin) that absorbs green and blue wavelengths and reflects red wavelengths of light.

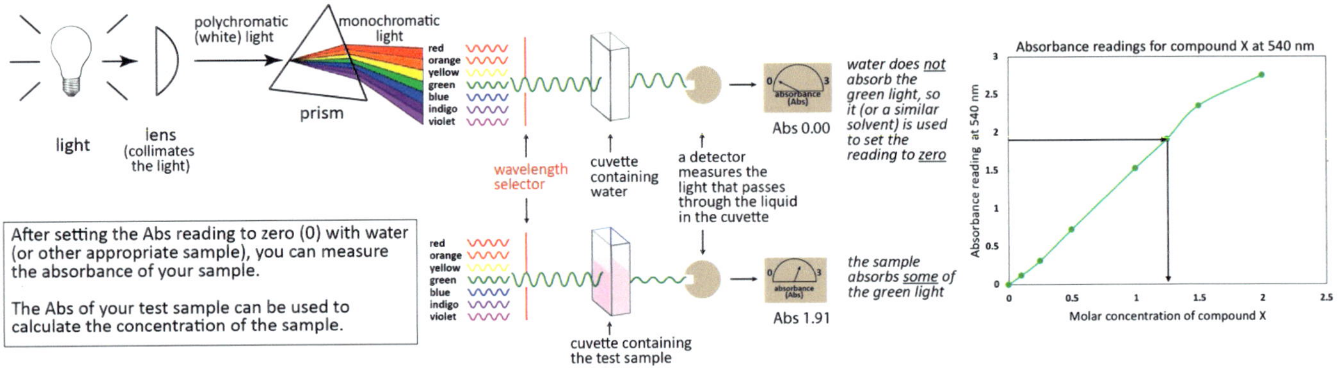

*Fig 3. The makings of a spectrophotometer (left). Light (tungsten for visible light) passed through a lens and a prism to separate the wavelengths of light. This makes it possible to look at whether a compound absorbs at all wavelengths or a specific wavelength, such as green (≈540 nm), shown here. In practice, you would set the spectrophotometer to **zero** (this is sometimes called BLANKING). Next (shown in the lower panel) you read the absorbance of your sample. The pink colored solution absorbs some of the green light. The detector measures the amount of light that is absorbed – in this case, the value is 1.91 which indicates that a significant amount of light was absorbed by the solution.*

The graph on the right shows absorbance of different concentrations of compound X (the test sample) at 540nm (green). The absorbance of the sample is linear from 0 to 1.5 M for this compound (X-axis). Notice that 0 M of compound X has an absorbance of 0; a 1.5 M solution of compound X has an absorbance of ≈2.4. From this graph, we can determine that the concentration of the sample in the cuvette shown on the left (Abs 1.91) was ≈1.25 M.

B. *Brief operating instructions for a typical spectrophotometer*

UV-visible Spectrophotometer 3100PC (VWR)

this instrument has a sample holder that can accommodate four cuvettes

Turn on machine (on/off switch in the back) and allow 15 min for the lamps to warm up.

↓

Insert a cuvette filled with water (or other solvent) into position #1 of the sample holder

(this will be the BLANK, which is used to set the absorbance to ZERO)

Insert cuvettes filled with your samples into position #2, #3, and #4

↓

Close the chamber

Select BASIC MODE

↓

Press: Go to λ

Use keypad to enter the desired wavelength

Press: ENTER

↓

The machine will BLANK (zero the reading) with the water in the position 1 cuvette

↓

Pull knob to bring cuvette #2 into the light beam

↓

READ: **Absorbance of sample in position #2**

Proceed to position #3 and #4

Practice 1:

Your absorbance readings for a blood sample (in triplicate) are as follows: 3.4, 3.9, 2.1

The patient is waiting for results. Should you use these values? Why or why not?

C. *Using the spectrophotometer to quantify chlorophyll concentration of a sample*

1. You need to calculate the concentration of chlorophyll in your plant samples to see if plant health is affected by an herbicide.
2. After looking at a spectral scan (wavelengths 400 to 750nm, Fig 4) of chlorophyll *a*, you decide to use the wavelength 665 nm (A_{665nm}) because chlorophyll *a* absorbs light at this wavelength. [You could also use 445 nm, but there is interference from other compounds that absorb light in this region.]
3. You obtain the following absorbance readings:

• Chlorophyll sample, undiluted	3.31
• Chlorophyll sample diluted 1:2	1.68
• Diluent only (alcohol)	0.0

4. To calculate the **concentration** of chlorophyll *a* you will use the A_{665nm} reading and the Lambert-Beer equation
   A = ε x L x c

For more information about the Lambert-Beer equation (also known as (aka) Beer's Law), visit https://phet.colorado.edu/sims/html/beers-law-lab/latest/beers-law-lab_en.html

A absorbance at the specified wavelength. The absorbance (A or Abs) is what you measure in the spectrophotometer.

ε extinction coefficient (or molar absorptivity). The extinction coefficient is a constant for a given compound at a 1M concentration in a defined solution (such as water or ethanol). Scientists have studied chlorophyll *a* extensively and have determined the **ε** value for this molecule. You can look up the **ε** value in the scientific literature and use it to easily determine the concentration of the chlorophyll that you prepared.

L length - the path length that light travels through your sample; usually 1 cm

c concentration of the solute (chlorophyll *a* in this example)

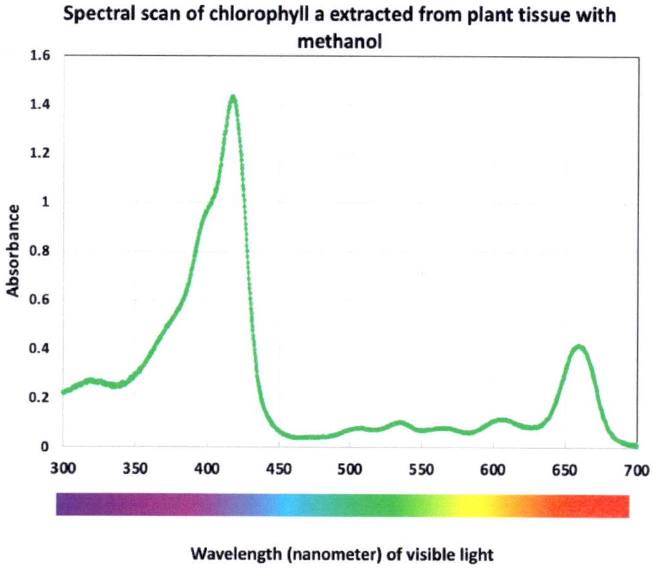

Fig 4. The visible spectrum of chlorophyll A (scanning spectrum). Each green dot represents the absorbance reading from the spectrophotometer at the wavelength indicated on the X-axis. The maximal absorbances occur at 419nm (absorbing violet light) and 660 nm (absorbing orange light). We see the color green when we look at chlorophyll a.
The absorption maxima for chlorophyll that has not been extracted from the plant using chemicals (which alter the chlorophyll structure) are 372 nm and 642 nm [Reference: B.F. Milne, Y. Toker, A. Rubio, and S. Brøndsted Nielsen (2015) Unraveling the Intrinsic Color of Chlorophyll, Angewandte Chemie International Edition 54 (7), 2170-2173]

The extinction coefficient (ε) of chlorophyll a depends on the wavelength and the solvent used to extract the chlorophyll.
ε = 84 $L\,g^{-1}\,cm^{-1}$ (this can also be written as $84\frac{liter}{g \bullet cm}$ at 665 nm for chlorophyll *a* dissolved in 95% ethanol. The value here is from appendix D in the book titled Chlorophyll by Einar Skarstad Egeland]
ε = 111,700 cm^{-1}/M at 417.8 nm (Strain, 1963)

5. Determine the chlorophyll (Chl *a*) concentration (in g/l) of a sample
   For this step, you need to review the Abs readings for your chlorophyll sample. The $A_{665\ nm}$ for the undiluted sample was 3.67. This value is out of the linear range for the VMR spectrophotometer (which is 0-3 absorbance units). You measured a second sample, in which the chlorophyll was diluted 1:2. The $A_{665\ nm}$ for the diluted sample was 1.68. Since this value is well within the acceptable range for the spectrophotometer, we will use this number, 1.68, to calculate the concentration of chlorophyll. Because this is a 1:2 dilution, you need to double this number to compensate for diluting: 1.68 x 2 = 3.36.

Remember:
$X^{-1} = \frac{1}{X}$

Rearrange Lambert-Beer to solve for **c**.

Concentration, c (g/l) = $A_{665\ nm}$/ε x liter

$$\text{Concentration, } c = \frac{\text{absorbance } A \text{ at chosen peak wavelength } \lambda}{\text{extinction coefficient } [L\,g^{-1}\,cm^{-1}] \bullet \text{ cuvette path length } [cm]}$$

The $A_{665\ nm}$ of your sample is 3.36. Use this value in the Lambert-Beer equation.

$$c = \frac{3.36}{84\ L\,g^{-1} \bullet cm^{-1}\ X\ 1\ cm} = 0.04\ g/l \text{ or } 40\ mg/l$$

D. *Using a spectrophotometer to determine the optical density (cell concentration) a culture*

Using the spectrophotometer to determine the density of cells in a liquid.

A spectrophotometer can also be used to determine the concentration of small particles in a solution, *e.g.*, microbial cells (Fig 5). Light with a wavelength of 600 nm is used to illuminate the sample. The detector measures the amount of 600 nm light remaining after the cells in the culture diffract the light. The resulting absorbance, A_{600nm} is often referred to as **optical density**, OD_{600nm} (or Abs_{600nm} or $A_{600\ nm}$). The 600 nm wavelength is used for microbial cultures because cellular compounds (proteins, DNA, chromophores) typically do not absorb at this wavelength and hence will not interfere with the measurement.

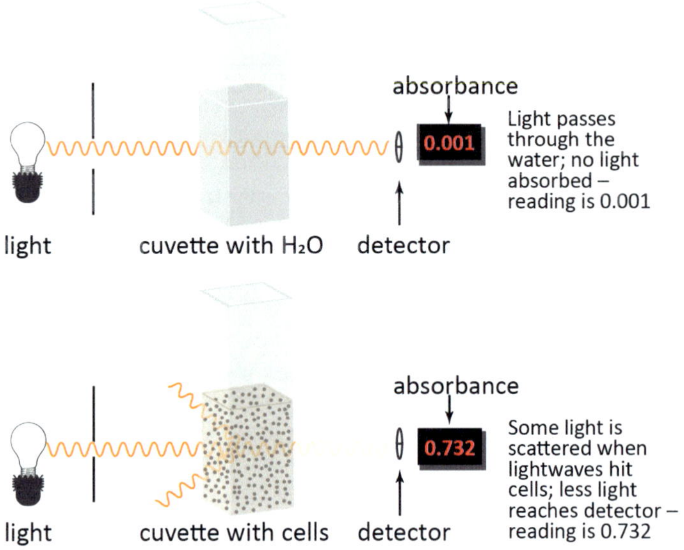

Fig 5. A spectrophotometer can be used to measure optical density.
Top: The illuminating wavelength of light (usually ≈600 nm) passed through a cuvette containing water because water does not absorb at 600 nm.
Bottom: Some light is scattered (diffracted) when it hits a microbial cell (the brown dots); scattered light does not reach the detector. A sample with more cells will diffract more light and the absorbance will increase. A standard curve can be created to determine cell concentration (cells/ml) based on the absorbance reading.

Uninoculated medium (shown in Fig 6) is clear (not cloudy); it does not absorb light with a wavelength of 600 nm, so all incoming light will reach the detector, A_{600nm} =0). After the clear, sterile growth medium is inoculated with live microbial cells, the turbidity of the medium increases as the cells divide. As stated above, the turbidity (cloudiness) of the culture liquid can be detected in the spectrophotometer because some of the 600 nm light hits the cells and is dispersed. The dispersed light does not reach the detector. - Since the detector measures less light, the absorbance of 600 nm, $A_{600\ nm}$ (or OD) increases.

Note that the spectrophotometer will give reliable readings on samples that are slightly to moderately turbid. Very turbid samples contain high numbers of cells (10^7 to 10^9). In very dilute cell cultures (or at high dilutions 10^{-6}, which is 1/1,000,000), there are too few cells to detect. The ultimate goal is to link absorbance to cell density (from the spot plate extrapolation). Hence, you will use simple dilutions of cells to get a reading in the spectrophotometer.

Fig 6. Growth medium. Left = uninoculated (sterile); right = inoculated medium after incubation (visibly cloudy or turbid).

If you know the OD or absorbance of a sample, you can determine the concentration of cells or compound, respectively.

Practice 2:
What two terms can be used to describe the output reading of a spectrophotometer when examining a cell culture at 600 nm?

_________________ and _________________

E. *Converting from optical density to concentration (number of cells/ml)*

For a scientist, it is useful to learn the relationship between cell number and optical density (OD). Establishing this relationship requires two separate measurements.

- First, you must measure absorbance of different dilutions of cells.
- Second, you must determine the number of viable cells in different dilutions using a plating assay.

The spectrophotometer has an upper and lower detection limit. Samples that are too dense or too dilute will not be measured accurately. The instrument is not sensitive enough for very dilute solutions of microbial cells *even though these solutions may contain up to a million cells/ml.*

In contrast, the plating assays are very effective at dilutions with very low cell concentrations, which then can be counted individually.

Hence, to combine the data from absorbance and plate counts, you need to compare two different types of dilutions: **simple dilutions** for the spectroscopy and **serial dilutions** for the plate counts. Sample readings for both types of dilutions are illustrated in Fig 7.

These estimates are based on growth of a microbial culture in rich medium after 1-3 days.

Spec readings	dilution	Example	5µl spot should contain:	plate counts
may be too high (> 2.5)	undilute	$5.0x10^8$	2.50E+06	TNTC
high	1:2.5	$2.5x10^8$	1.25E+06	TNTC
med	1:5	$1.0x10^8$	5.00E+05	TNTC
med	1:7.5	$6.6x10^7$	3.30E+05	TNTC
med	1:10	$5.0x10^7$	2.50E+05	TNTC
low	1:20	$2.5x10^7$	1.25E+05	TNTC
low	1:50	$1.0x10^6$	5.00E+03	TNTC
low, 0	1:100	$5.0x10^6$	2.50E+04	TNTC
0	1:1000	$5.0x10^5$	2.50E+03	TNTC
0	1:10,000	$5.0x10^4$	2.50E+02	250 (still TNTC)
0	1:100,000	$5.0x10^3$	2.50E+01	25 (still too many)
0	1:1,000,000	$5.0x10^2$	2.50E+00	2.5 (≈2-3 colonies)

TNTC = too numerous to count

[This example is for a hypothetical organism with an initial density of 5 $x10^8$ cells/ml]

Fig 7. Cells must be diluted prior to reading absorbance in the spectrophotometer and before spot plating. Note that different dilutions are needed for each of these assays if you want to obtain useable results.

F. *Using plate counts and spectroscopy to create a standard curve to quantify cell density*

This section illustrates how you can use A_{600} *and* cell concentration (from the spot plate in the previous primer) to generate a standard curve (standard graph) which serves as your permanent reference. You only need to make a standard graph for a particular organism ONCE. Thereafter, you simple read the absorbance of a culture and refer to your standard curve to determine the cell concentration.

> Ideally, a standard curve is should be a straight line (not a curve) if your calculations are correct, bench work is accurate, and sample dilutions are within the linear range of the spectrophotometer.

Creating a standard curve — a reference for future use

EXAMPLE

- You prepare simple dilutions of a culture of cells identified as *Xancromagnus* (a fictitious organism).
- You use dilutions similar as those in Fig 3 that will yield spectrophotometer readings and follow the instructions for reading absorbance at 600 nm (A_{600}) on the spectrophotometer.
- You record the A_{600} for each dilution. The data are shown in Table 1 below. These are hypothetical data; you will have more data points than shown here. Note that the A_{600} for the undiluted sample, 3.5, is quite high. The upper limit for the spectrophotometer is ≈ 2.5 – 3 absorbance units. Therefore, you should not use this data point in your graph.

Table 1

Dilution	A_{600nm}
Undilute	3.5
1:2	1.8
1:5	0.88
1:10	0.38

Part 1: Gather the data

	Estimated cells/ml in different dilutions derived from the cell culture*	Absorbance
Undiluted culture of cells	1×10^8 cells/ml (from the spot plate)	3.5 (too high to use)
1:2 or $\frac{1}{2}$	5×10^7 cells/ml	1.8
1:5 or $\frac{1}{5}$	2×10^7 cells/ml	0.88
1:10 or $\frac{1}{10}$ or 10^{-1}	1×10^7 cells/ml	0.38

*Practice data - The original culture contains 1×10^8 cells/ml, which you obtained previously from spot plating aliquots of serial dilutions.
Hence a 1:2 dilution contains 1×10^8 cells/ml $\div 2 = 0.5 \times 10^8$ cells/ml or 5×10^7 cells/ml
Similarly, the 1:5 dilution contains 1×10^8 cells/ml $\div 5 = 0.2 \times 10^8$ cells/ml or 2×10^7 cells/ml

Part 2. Use the absorbance data in red (above) to create your standard graph

Here are the steps we took to create the graph shown in Fig 8.

1. We put the data (cells/ml and absorbance) into Excel
2. We selected data and picked the scatter plot (Fig 8A)
3. We added axis labels and title
4. We changed the numbers on the horizontal axis to scientific (Fig 8B)
5. We added a trendline (you can access this by clicking on the blue data points in the graph; to display the R-value — see Format Trendline)

Cells/ml	A 600
5.00E+07	1.8
2.00E+07	0.88
1.00E+07	0.38

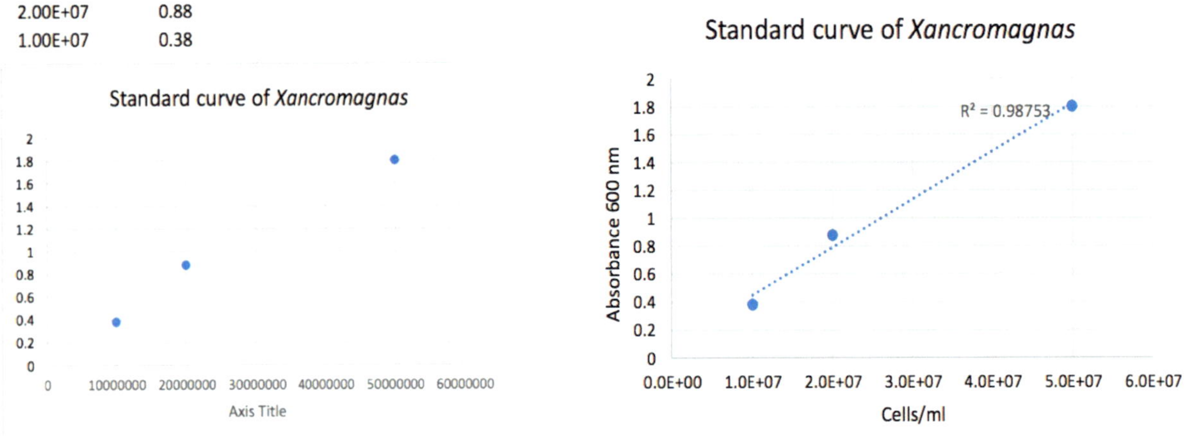

Fig 8A. A scatter plot shown on the left was created using the raw data shown above this graph. The numbers were selected in the Excel file and a scatter plot option was used. This is the initial (crude). The improved version of the graph on the right. Notice that both the X and Y axis have been labeled and the X axis numbers have been converted to scientific notation. The data points were selected to create a Trendline (blue dashed) and report statistical information (R).

Practice 3:

You plated 50 µl of a 10^{-6} dilution of your bacterial cell culture. After 3 days at room temperature incubation, you count 183 colonies on the plate.

1. How many cells were in the 50 µl aliquot of the 10^{-6} dilution? ___________ cells
2. How many cells do you expect in 1 ml of the 10^{-6} dilution? ___________ cells
3. Finally, 1 ml of the undiluted cell culture should have approx. __________ cells/ml

G. *Troubleshooting*

Problem	Possible reason	Possible solution
I cannot get a zero reading for the water control (the 'blank')	You forgot to press the ENTER or did not put the machine in Basic Mode	Look at the detailed photos in the Methods Manual
My sample from a serial dilution of bacterial cells is reading -0.01.	You may be trying to read a sample that is too dilute. The instrument is not sensitive to very dilute samples.	Use a more concentrated sample to read absorbance
The absorbance readings for my dilutions are not linear	You might have made an error in pipetting or calculation Sample may be too concentrated or too dilute (not in the linear range of instrument)	Check your calculations If one sample is near the limit (Abs =3), you may want to ignore this reading
My readings make no sense or the values are erratic	The cuvette may be in the wrong orientation	Make sure the clear part of the cuvette is in line with the light path
	Perhaps you put an Eppendorf tube in the spectrophotometer chamber	Transfer liquid to the cuvette.
I cannot get a reading for my samples	The spectrophotometer may not be in the right mode	Restart the initiation sequence
	The light bulb may have burned out,	The reading for the lamp (on the top of the screen) should be higher than 20,000
My absorbance readings are very low	You may not have set the instrument to read at the proper wavelength	Check that you are using the correct wavelength

13. FIGURES (GRAPHS, TABLES, AND PHOTOS) COMMUNICATE YOUR RESULTS

In an experiment, the investigator chooses the independent variable and makes observations or takes measurements (dependent variable). After collection of the raw data the investigator will prepare tables, photos, and graphs that communicate the data.

This primer is designed to help you understand *which type of graph* is appropriate for the data that have been collected. The graph must be labelled properly and should include a figure legend that will help the reader understand what was measured, how measurements were taken, and how data were evaluated. The choice of graph depends on the type of the data collected.

COMMON GRAPH STYLES:

- **line graph** consists of a horizontal and a vertical axis. Individual data points are marked on the graph and connected by a line. Thus, a line graph is most useful for data involving a continuum (a continuous sequence, *e.g.,* changes over time). The local change from point to point can be seen. The line connecting the individual data points is often not straight.
- **scatter plot** is similar to a line graph in that it also consists of a horizontal and a vertical axis and individual data points are marked on the graph. The data points are not connected by a line. Scatter plots are used to determine correlations and show how much one variable is affected by another variable. A trend line and the R^2 best fit values can be used to show the level of correlation, (an R^2 close to 1 indicates a high level of correlation).
- **bar graph** is more appropriate for categorical data, where discrete categories that are being compared. One axis (most often the y-axis) shows the measured variable while the different categories are indicated on the other axis.
- **histogram** is a type of bar graph the shows the distribution of numerical, but "binned" data. Individual measurements (data points) are organized into intervals that are adjacent to each other. The intervals are usually the same size.
- **pie chart** is used to illustrate numerical proportions with the entire "pie" representing the whole, *e.g.,*100%.

Important parts of all graphs

Title	The title of a graph describes the content of the graph. It is helpful to give the "take-home message"
Axis labels	The X-axis (horizontal) and the Y-axis (vertical) are labeled with the type of information, including units that are graphed
Scale	The scales are the numbers placed along the X and Y axes; units might be time (hr, sec, min), distance, cell number, absorbance, enzyme activity, *etc.* Bar graphs have categories on one axis (often the X-axis)
Data points and error bars	Data points are noted in line graphs and scatter plots; it is good to add error bars to all graphs
Data labels	The labels should be clear and concise
Lines	In a line graph the data points are connected in the graph; the line may be rough or smooth. For scatter plots programs such as Excel can create trend lines including statistical analyses how well the data fit the trendline
Figure legend	Each graph should have a figure legend that describes to the reader what is shown in the graph without having to read the whole report

A. *Line graphs*

A line graph is used for data involving a continuum, such as time, temperature, concentration, in which data points are connected by a line. Note that the line does not need to be straight (as shown in the figure here), but the points are connected because the data points are part of a sequence (continuum). Fig 1 is a line graph created from data from Table 1 (concentration of copper sulfate on the X axis and copepod survival on the Y axis) .

Table 1: Copepod survival

$[CuSO_4]$ mM	copepod survival (n= 20)
0	20
0.1	20
0.5	19
1.0	18
1.5	15
2.0	9
2.5	3

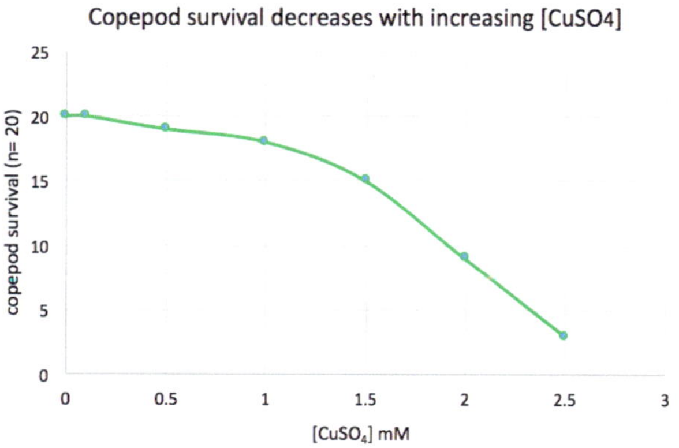

Fig. 1 A line graph showing copepod survival as a function of copper sulfate concentration.

Practice 1:

1. Why are line graphs useful?

2. Use the line graph above to determine the number of copepods that are expected to survive if the concentration of $CuSO_4$ is 1.75 mM.

EXAMPLE

You have done a chemical analysis to determine the percent of sugar in various foods. The results are shown in Table 2.
For a public presentation, it is easier for the audience to compare these values by looking at a graph, so you create a line graph as shown below (left panel).

Table 2 Sugar content of common foods

Food item	Sugar (g/100 g of food item)
ketchup	8.8
peanut butter	9.2
chocolate bar	33.2
ice cream	21.4
chocolate cake	30.3
Pepsi	28.9
cracker	11.8

Why this does not work

The first line graph shows food items (X-axis) arranged according to sugar content (% sugar). While this graph *looks reasonable*, a line graph (left) is not the best way to visualize these data because other than sugar, there is not relationship between the different items. The line connecting the food items implies that they are related as part of a continuum. Notice that if the graph is created based on alphabetically-arranged food items, the line generated looks completely different. Since the food items *are not related*, the preferred type of graph is a bar graph, shown in Fig 3.

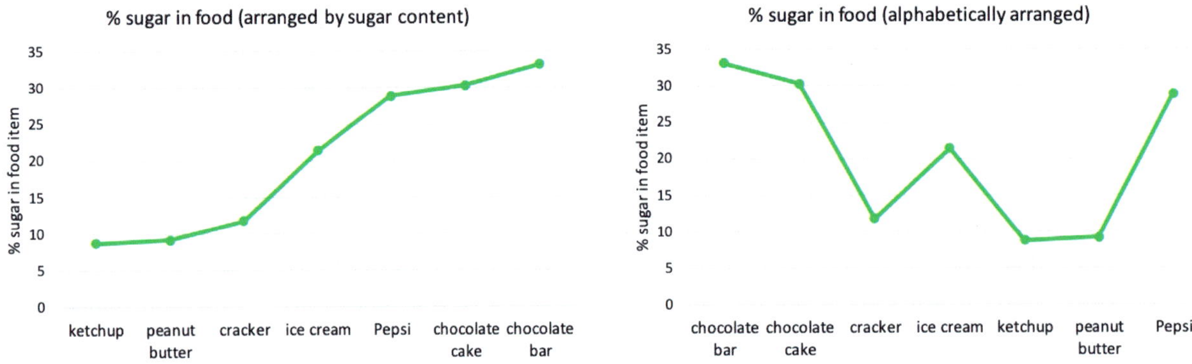

Fig 2. The data in table 2 were rearranged to create two different line graphs. This type of information (unrelated food items) should NOT be presented as a line graph.

B. *Bar graph*

A bar graph shows the relationship between two variables and is most appropriate when the independent variables comprise distinct categories (*ie.* not a continuum, such as time). The choice of bar graph rather than a line graph is important because these food items are unrelated to one another except for the fact that they contain sugar. Since there is no continuum, so a bar graph is the best option here.

Why are bar graphs useful?

Bar graphs give the reader/audience a visual mechanism for quickly interpreting data. For example, in the graph above, one can "see" that a chocolate bar has roughly 3 times (3X) more sugar than ketchup or peanut butter. Bar graphs are not useful for *predicting* information as described for the line graph.

A **bar graph** (Fig 3) is the appropriate type of graph for visualizing categorical data.

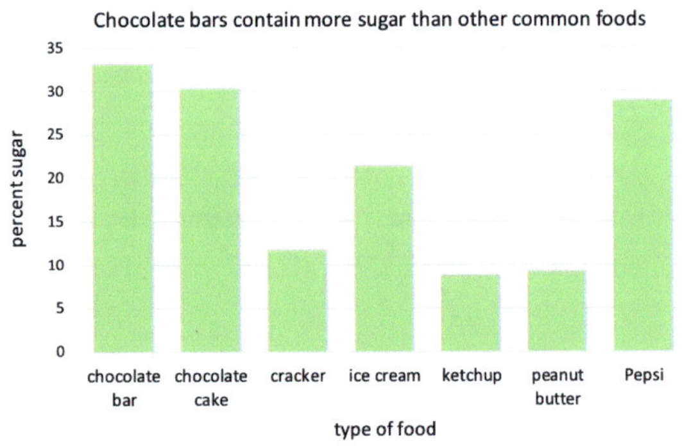

Fig 3. A bar graph is the preferred type of graph to show a relationship between unrelated items (such as food items)

C. *Scatter plot*

A scatter plot is used to determine if a correlation exists. You would use this type of graph if, for example, you wanted to know if students who completed the practice homework finished the class with an overall higher final grade. Some raw data are shown in the table below and these numbers were used to generate a scatter plot in Excel. Examine the graph shown below — do you think there is a correlation between the number of homework assignments completed and final grade?

Student	Final score (out of 100)	# of completed assignments
CF	98	9
LH	92	10
SS	95	8
PL	90	9
FJ	93	9.5
AG	85	6
RP	82	7
Yb	73	3
NR	77	1
GRN	75	5
RC	71	2
WB	68	1
ID	64	4

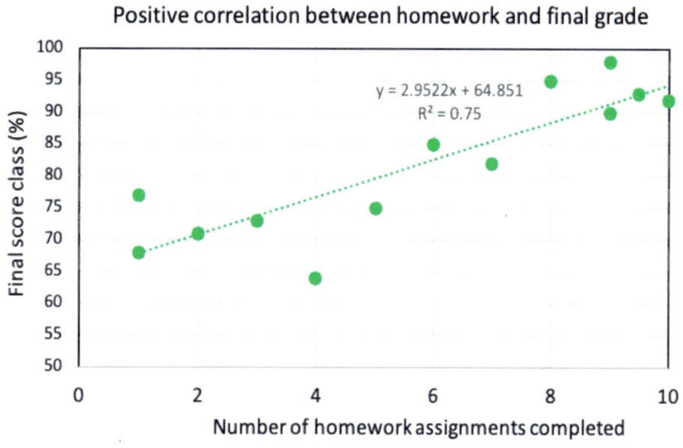

Fig 4. This scatter plot shows a correlation between homework completed and final grade.

D. *Histogram*

A histogram is a type of column chart that depicts binned (grouped) items. It is used to visualize the distribution of numerical data and frequency.

EXAMPLE

You have been looking at genes in a particular organism. From a segment of the genome, you determine the lengths of genes in numbers of nucleotides/gene. Your results are as follows: the number of nucleotides in genes 1-31 is: 10647, 483, 933, 588, 5133, 876, 2073, 2313, 1938, 792, 1026, 4062, 2820, 3412, 936, 4887, 13713, 8652, 1578, 12864, 1485, 9132, 7884, 15924, 5667, 4557, 6222, 2621, 10008, 7239, and 12102.

These numbers are not very useful, but if you enter the numbers into Excel (or similar software) and select the histogram option, the software (algorithm) will bin (group) the genes mathematically — by size (number of nucleotides) in this example. The bin size for the numbers listed above is 2000. There are 11 genes that range in size from 500 to 2500 nucleotides (588, 792, 876, 933, 936, 1026, 1485, 1578, 1938, 2073, and 2313). The next bin contains four genes between 2501 and 4500.

A histogram generated from the numbers listed above is shown in Fig 5.

The height of the bars a histogram represents the relative size of the bin. In this example each bar represents the number of genes in a particular size range, listed in the x axis. The data are arranged by increasing size (number of nucleotides). When arranged this way it is easy to see that 'small' genes are more common than 'large' genes *in this organism*.

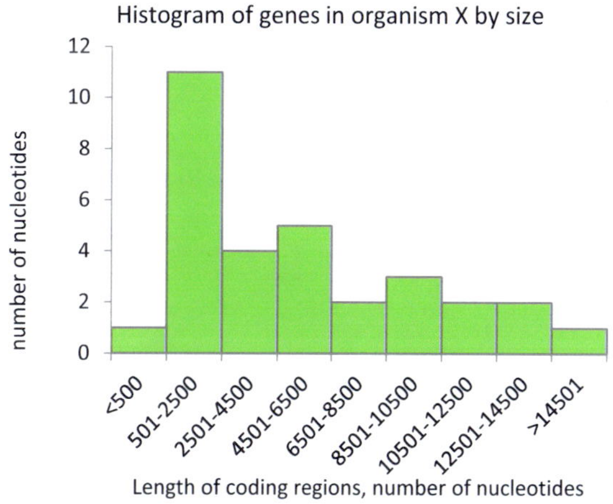

Fig 5. A histogram bins (groups) items that are related numerically.

E. *Pie charts*

A pie chart is useful for visually displaying data that represent parts of a whole (such as 100%). Each of the categories (measurements) are sized as the respective fractions adding up to one whole.

Cara makes $12.50/hr and works about 60 hr/month. She pays about $40/month in taxes. her take-home pay is ≈**$710** a month (after taxes). This number represents the whole.

The table below lists her known monthly expenses, an estimate of food costs and leftover amount (disposable income) for entertainment, clothing etc. The pie chart in Fig 6 provides a visual representation of her monthly expenses.

Monthly expense	Spending amount
Rent	$248
Car insurance and gas	$96
Phone	$72
Cable	$37
Savings	$40
Food	$115
Health care (co-pay)	$25
Utilities	$41
Disposable	$36
TOTAL (the WHOLE)	$710.00

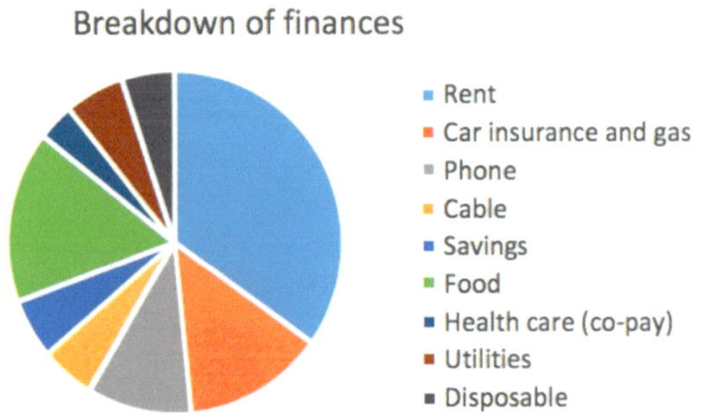

Fig 6. Cara's budget. In this example, Cara looks at her total income (100%) and aliquots fractions of her income to the different categories each month.

Practice 2:
Examine the daily temperature maxima, shown in the table below. Which type of graph, a line graph (Graph 1) or a bar graph (Graph 2), do you think best represents the data?

Temperature highs in Moscow

Day	Date	Temperature
1	April 6	15° C
2	April 7	16° C
3	April 8	15° C
4	April 9	20° C
5	April 10	21° C
6	April 11	22° C

Graph 1

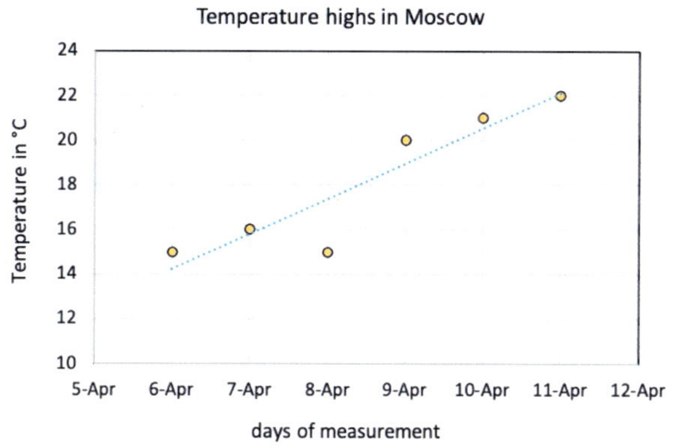

Graph 2

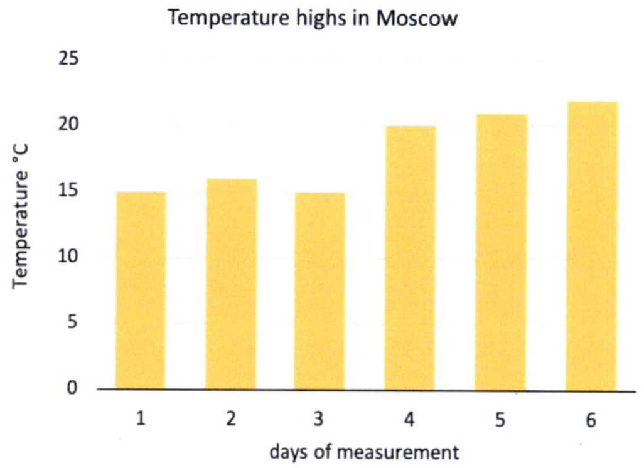

Fig. 7 Optional graphs to illustrate graphing data
Graph 1 is an XY Scatter plot; the blue line was created using the Trendline tool (in Microsoft Excel)
Graph 2 is a column bar graph created using the same data.

Practice 3:
You designed an experiment to determine the optimal pH for growing a new strain of yeast that you isolated from an apple. The experiment was done in triplicate (trials 1-3). Yeast cells were grown to 5×10^8 cells/ml, serially diluted to 5×10^3 cells/ml and 50 µl aliquots of the dilution were spread on agar medium at the pH listed in column 1 of the table. After 2 days, the number of colonies were counted. Look for outliers (numbers that are very different than their cohorts), then calculate the average value for each pH and graph the results.

	Number of colonies			
pH of the agar growth medium	**trial 1**	**trial 2**	**trial 3**	**Average**
3.5	120	98	114	
4.0	140	112	122	
4.5	171	162	155	
5.0	211	190	225	
5.5	177	185	191	
6.0	145	138	152	
6.5	132	128	26	
7.0	51	48	62	
7.5	34	40	41	

F. *Graphing do's and don'ts*

Problem 1: Incorrect spacing of units on the X and/or Y axis.

Incorrect spacing of units is a *major* problem because it misrepresents the data if the unit spacing is not correct. The spacing between units must be uniform and justified. Look at the data in Table 1. These data are represented in the four different graphs below.

Table 1

Soil depth (cm)	Number of microbial cells/g of soil
0	1×10^5
5	9×10^5
10	3×10^6
20	2×10^7
40	7×10^6
100	5×10^5

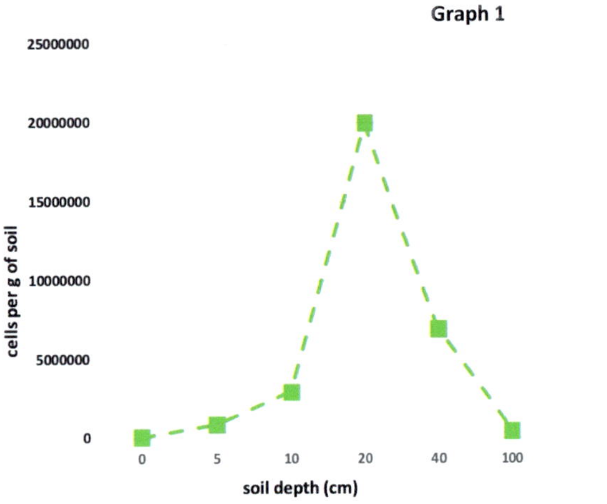

Fig 6. Graph 1: this shows a typical line graph that is generated in Excel. To more accurately represent the number of cells at each depth, the scale for each axis needs to be modified.

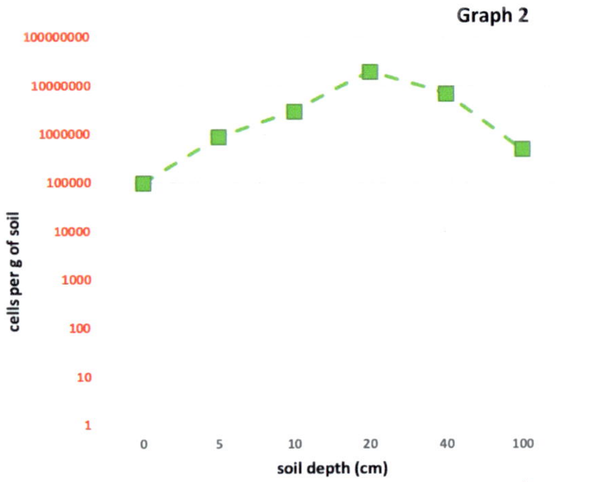

Fig 7. The first improvement is shown in Graph 2. The data on the y-axis show the number of microbial cells per gram of soil. The y-axis has been set to a log scale to reflect the actual cell concentration. The x-axis is still incorrect. The numbers are not justified and give a distorted sense of the soil depth.

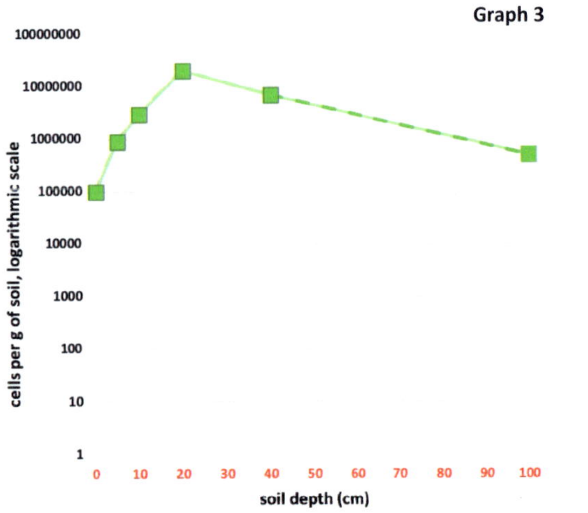

Fig 8. The second improvement in shown in Graph 3. The spacing for the x-axis has been justified. This gives a realistic view of the numbers of microbes in soil as a function of depth.

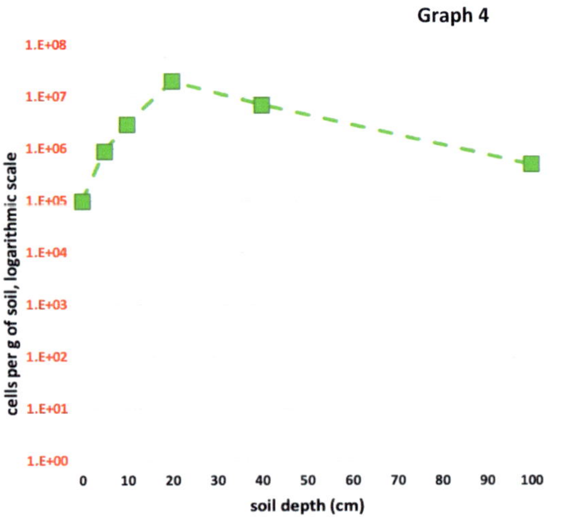

Fig 9. Graph 4: the only difference between graphs 3 and 4 is that the units for the y-axis are given in scientific notation in graph 4. This makes it easier for the viewer to read and interpret the numbers.

How to determine the spacing to justify numbers

	Example (from the data above)
First: Identify the range.	soil depths: 0, 5, 10, 20, 40, 100 cm
Second: Measure the spread	Spread = 0 cm to 100 cm = 100 cm

Third: Distribute the spread evenly

Look at the graph on the right showing soil depth; notice that the spacing is uniform and uses as much of the x-axis as possible.

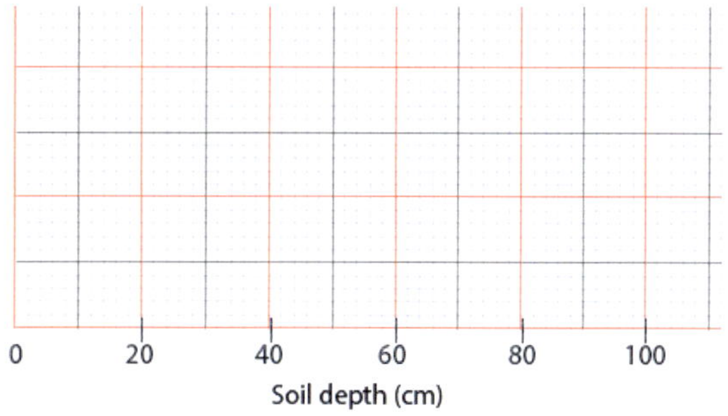

Problem 2: Insufficient labeling of graphs

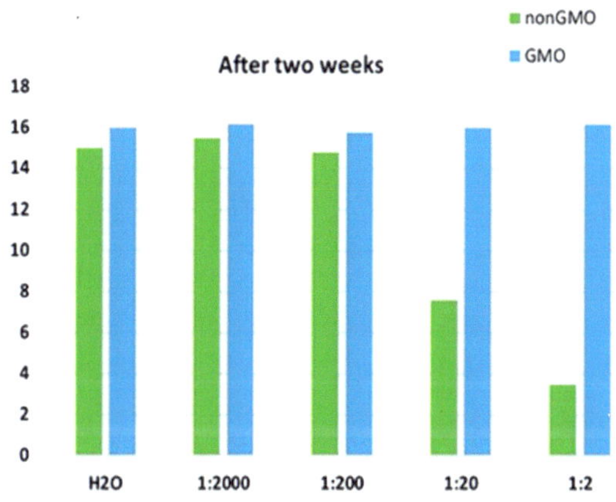

Fig 10. It is essential to have correct labels on a graph. This graph is poorly labeled — no y-axis label, nothing to define the numbers on the x-axis, and a title that gives only partial information. The reader cannot possibly decipher the information.

The graph to the left is lacking several critical pieces of information

- First, the title is vague.
- Second, the y-axis label is missing. It has been added in the graph to the right – information about plant height (cm) and the number of plants per treatment (N=3).
- Third, we do not know how to interpret the x-axis values. Providing a dilution value is insufficient information since we do not know what was diluted or the original concentration of the compound being tested.

An improved version of the graph above is shown on the right.

- An informative title has been added.
- The y-axis label has been added – information about plant height (cm) and the number of plants per treatment (N=3).
- The x-axis values have been corrected. The name of the chemical being tested (Roundup) was added and the percentage of Roundup (the concentration) used in each of the tests was calculated. Example: the original concentration of Roundup was 8%. A 1:20 dilution resulted in a concentration of 0.4% Roundup.

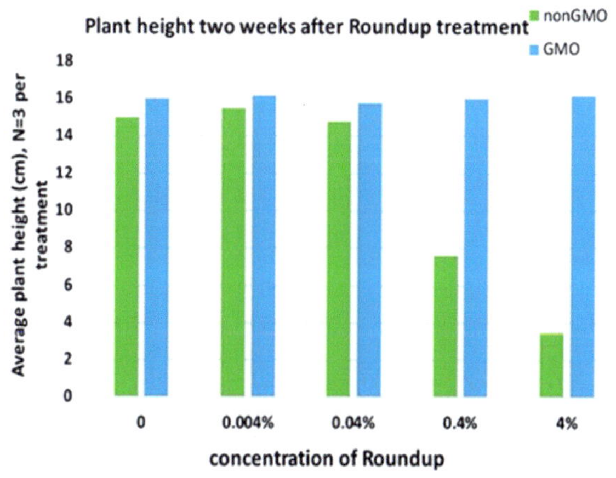

Fig 11. This graph has helpful labels. The concentration of Roundup on the x-axis communicates accurate information to the reader. The y-axis tells the reader what is being measured.

14. COMMUNICATE YOUR WORK IN WRITING

A. *Critical elements for a manuscript*

General thoughts

- The descriptions for manuscript sections below are designed for an undergraduate student research project – a mini-manuscript
- Use clear sentence structure, concise (no circumlocution), and objective (non-biased) language (no emotional attachment to your results)
- Most journals still use third person: for example, rather than saying "we believe that …", you might say "the data suggest that…")
- Check spelling and grammar
- Include the Latin names for plants and organisms in *italics*
- Use the **4C writing method for at least three different types of sentences** (see notes below)
- Use 3-5 figures to tell a story about your work; do **not** use raw data in your figures

---

Title of the study

- Provide a title that gives a brief result – for example, instead of "Role of glyphosate on GMO alfalfa", you could use " Glyphosate induces a stress response in GMO alfalfa"
- After the title, provide list of key authors, department and home institution

---

Abstract ≈200 words

Components of the abstract include:

- General information about the topic (1-2 sentences)
- Specific information related to your topic that introduces the reader to the **significance** of the problem
- An introduction to the topic as it relates to the scientific question in the manuscript
- A statement of the hypothesis and/or the aim(s) of the work
- A summary of results; provide some detail (*ie.* state the number of trials and samples tested)
- Conclusions

Background, literature review ≈400-600 words

3-5 references from primary literature – not websites

Components that belong in the background, literature review include:

- **General information** about your topic which you will find by reading the original scientific literature (start by searching PUBMed). For example, if you are studying an agricultural plant, you may want to describe what is known about the plant (SIGNIFICANCE FOR SCIENCE) and explain why it is an important crop (SIGNIFICANCE FOR SOCIETY).
- More specific information about the project. A few (2-4) sentences that define the question as it relates to this project and the problem that you are studying in general terms.
- Data from recent, relevant publications
- Be sure to identify the gap in knowledge (this should tie-in with the work that you are doing for your IRP).
- Summarize information from your project that may reduce the gap in knowledge.

Methods ≈500-800 words

Components that belong in the Methods include:

DO NOT DISCUSS RESULTS IN THE METHODS SECTION

- Information about all organisms (complete Latin genus species names) that were used in the study, *i.e.* microbial strains, animals, or plants used in the study, how they were grown, and the media used for growth.
- All reagents used, including stock solutions and buffers. If dilutions were made, describe how dilutions were made (volumes *etc.*).
- Describe when and how the experiments were done, include information on controls
- Lists of materials that you used *do not* belong in a manuscript

---

Results ≈500-600 words

3-4 figures/tables with titles and legends: they will be submitted figures (photos, graphs, tables) separately during the semester

> Be sure to calculate the average values for your samples before you prepare figures and tables – DO NOT provide raw data or individual values for each experiment.

Components that belong in the Results include:

- A statement that reminds the reader about the goal of the project.
- A description of results in a logical progression. Chronological order is often a reasonable way to present the results.
- Use the 4C annotation method (**C**ALCULATION, **C**ONTEXT, **C**OMPARISON, **C**LARITY) to construct at least **THREE different types of sentences in the results section. Each of the 4C elements must be within a single sentence; see information at the end of this guideline section.**
- Properly labeled figures and tables arranged in a sequence that makes sense.
    - Refer to each Fig/Table in the text. Look at each Fig/Table and write text that describes what is shown.
    - To add a caption, press control and click on the figure, and scroll down to insert caption. Use the GROUP function to attach the caption to the figure so that they move together.

- **Wrap text around the figure+ caption** (see instructions at the end of this section).

Discussion ≈6-8 sentences

Use a few simple, declarative sentences to explain the meaning/significance of the experimental findings. In the discussion, you must interpret your results and explain how your results align with established information about the topic. Relate your results to your original hypothesis and to the published literature. Explain whether the data support or refute your hypothesis. You should **not repeat** the information that you have written in the results, but you can provide further analysis. For example, if you show a bar graph figure in Results, you can discuss the significance of different values that you obtained.

1. Restate (in fresh words) your goal(s), scientific question, and/or hypothesis
2. Describe result #1
3. Comment about the significance of result #1 (impact on society *etc.*)
4. Describe result #2
5. Comment about the significance of result #2
6. Relate your findings to the published literature
7. Discuss limitations or contradictions of your findings (optional)
8. Describe the major impact of your results

References

The best source for searchable, peer-reviewed articles is PUBMED https://www.ncbi.nlm.nih.gov/pubmed/

General format for the IRP:

- In the text: [#] (Number sequentially, *i.e.* [1], [2]) or (author last name, year of publication); if the author name is included in your sentence, include the year of publication in parentheses.
- Reference list at the end of the manuscript should follow this format: 1st author first initial, last name, 2nd author first initial and last name, 3rd author first initial and last name and last author first initial last name (year of publication in parenthesis) Title of publication. *Journal title* (in italics) **volume (in bold)**: pages of article

EXAMPLE

A. Wahid, S. Gelani, M. Ashraf and M. R. Foolad, (2007) Heat Tolerance in Plants: An Overview. *Environmental and Experimental Botany*, **61**:199-223.

Note that some journals, such as PLoS (Public Library of Science) are published online only and may lack page numbers — in this case, you can provide the information shown in **BOLD** below:

EXAMPLE

S. Park and M.S. Kim (2018) An event-related potential study of spatial working memory in binge drinking college students. PLoS One **13(9):e0203696. doi: 10.1371/journal.pone.0203696**

B. *Formatting techniques*

Wrap text around a figure in Word

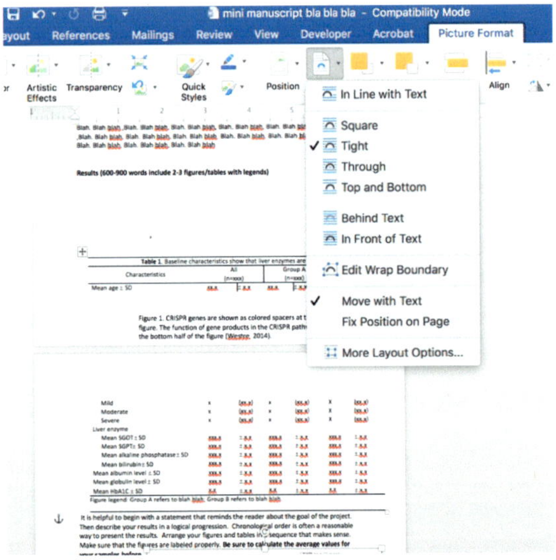

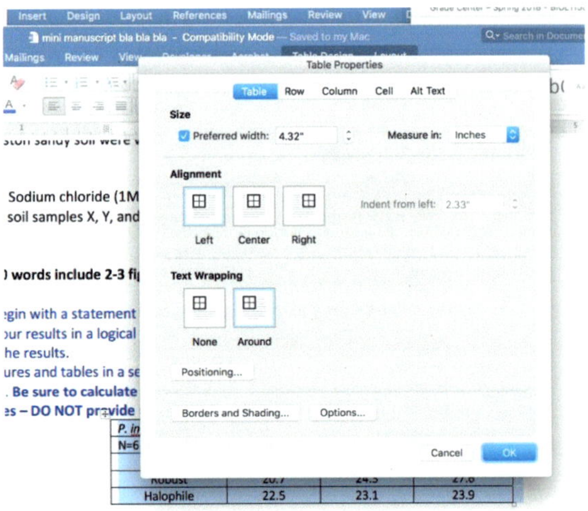

Wrap text around a Table in Word.

Select the table, go to the Table menu and select Table Properties.

Examples of descriptions found in a Methods section:

Strains and growth conditions

List Genus species, the type of medium (LB for *E. coli*, YPD for yeast etc.), temperature (*e.g.,* room temperature, 30°C, 37°C) and condition (on a rotator wheel, shaking *vs* static, light or dark, *etc.*).

EXAMPLE

Planta indeterminae varieties included cv. Halophile, Sensitive, Robust, and Wimpy. *P. indeterminae* seeds were germinated in Petri dishes. Germination rate was calculated as X% germination after seven days. Seeds were planted in 1.5-inch pots containing washed sand. The pots were maintained at 24°C in the greenhouse with daily watering.

Preparation of medium (soil, *etc.*)

EXAMPLE

Four soil samples from Jensen lawn (A), field dirt (give GPS location) (B), commercial soil (C; include the brand name) and Lewiston sandy soil (D;GPS XX.XX) were soaked in distilled H_2O for 20 min. Excess water was drained and damp soil was packed into 12, 3x3 inch plastic plant pots

TREATMENTS

1 M NaCl (5.84 g/100 ml) was diluted to __ M with distilled H_2O and ____ ml volume was added to soil samples X, Y, and Z. An equal volume of distilled H_2O was added to the no-salt control.

Raw data versus final information for a report/mini-manuscript

Six seedlings of each plant type were measured (in cm) 40 days after germination

Cultivar:	***P. indeterminae* cv treatments**		
	H_2O	**5 mM NaCl**	**50 mM NaCl**
Wimpy	24.6, 25.1, 25.4, 23.8, 25.8, 24,4	28.4, 29.3, 29.1, 27.6, 26.8, 25.3	0, 0, 0, 0, 0, 0
Robust	20.7, 19.5, 21.2, 21.7, 23.1, 19.9	24.6, 23.1, 25.1, 24.3, 23.9, 24.6	27.6, 27.9, 26.4, 26.1, 26.5, 25.4
Halophile	24.2, 21.8, 22.6, 22.1, 23.1, 22.7	23.4, 24.5, 21.9, 23.3, 24.8, 25.1	22.5, 23,8, 23.4,22,9, 23.7, 24.7

Below: The mini-manuscript version of the table containing the average values: this is the information that you publish.

***P. indeterminae* cv**	**Average plant height 40 days after germination**		
N=6 for each type	**H_2O**	**5 mM NaCl**	**50 mM NaCl**
Wimpy	25.4	28.4	0
Robust	20.7	24.3	27.6
Halophile	22.5	23.1	23.9

C. *Include 4C content (Calculation, Context, Comparison, Clarity) to write sentences for the Results section*

- a Calculation refers to a number derived from a calculation related to your ***results.*** The concentration of one of your solutions does not qualify as a calculation.
- information about the Context of the results
- a Comparison that is being made (*i.e.,* comparing a sample with the control)
- written with Clarity – no extraneous information (is the sentence **C**lear?)

Reference: Ruscetti T, Krueger K, Sabatier C (2018) Improving quantitative writing one sentence at a time. PLoS ONE 13(9): e0203109. https://doi.org/10.1371/journal.pone.0203109

TIP: Reading your sentences out loud will help you detect spelling/grammar mistakes and determine whether your organization is logical and the descriptions make sense.

15. INDEX

Made in United States
Troutdale, OR
11/03/2026

56844387R00043